Meat and Nutrition

Edited by Chhabi Lal Ranabhat

Published in London, United Kingdom

IntechOpen

Supporting open minds since 2005

Meat and Nutrition
http://dx.doi.org/10.5772/intechopen.78869
Edited by Chhabi Lal Ranabhat

Contributors
Jag Pal, Om Pravesh Kumar Ravi, Sangeeta Kumari, Akhilesh Kumar Singh, Yuanlong Cui, Saffa Riffat, Xuan Xue, Raffaela Biesuz, Lisa Rita Magnaghi, Adrian C. Williams, Lisa J. Hill, Eduardo Eugenio Spers, Thelma Lucchese, Pedro Carvaho Burnier, Paolo Polidori, Silvia Vincenzetti, Paola Di Girolami, Dhary Alewy Almashhadany, Almudena Soriano, Carlos Sánchez-García, Bruno I. Cappellozza, Rodrigo S. Marques, Krishna Prasad Pathak, Emanuela Mattos, Edward C. Webb, Manuel Eber E. Paredes Arana

First published in London, United Kingdom, 2021 by IntechOpen
IntechOpen is the global imprint of INTECHOPEN LIMITED, registered in England and Wales, registration number: 11086078, 5 Princes Gate Court, London, SW7 2QJ, United Kingdom
Printed in Croatia

British Library Cataloguing-in-Publication Data
A catalogue record for this book is available from the British Library

Additional hard and PDF copies can be obtained from orders@intechopen.com

Meat and Nutrition
Edited by Chhabi Lal Ranabhat
p. cm.
Print ISBN 978-1-83968-702-0
Online ISBN 978-1-83968-703-7
eBook (PDF) ISBN 978-1-83968-704-4

For EU product safety concerns:
IN TECH d.o.o., Prolaz Marije Krucifikse Kozulić 3, 51000 Rijeka, Croatia,
info@intechopen.com or visit our website at intechopen.com.

Meet the editor

Dr. Chhabi Ranabhat is a Research Scientist at the Global Center for Research and Development and a previous research fellow in the Policy Research Institute, Nepal, and a professional researcher at the Institute for Poverty Alleviation and International Development, Yonsei University, Republic of Korea. He earned his Ph.D. in Health Service in 2016 from Yonsei University and special fellowship training on the global burden of disease and policy implication from the Institute for Health Metrics and Evaluation, University of Washington. He received two master's degrees, one in Public Health from BP Koirala Institute of Health Science, the top health science university in Nepal, and another one in Sociology from Tribhuvan University, Nepal as an outstanding student. He has more than seven years of working experience in the health system of Nepal as a policy expert, about two years of experience in World Health Organization Nepal research projects, and about three years of experience in Good Neighbors International, Nepal. His area of expertise is Health Service and Policies regarding The Global Burden of Diseases. He has published more than 60 research papers (https://scholar.google.com/citations?user=JKDGO8oAAAAJ&hl=en) in reputed journals and is the editor and reviewer of scientific journals with impact factors.

Contents

Preface

Essential amino acids, the majority of which come from consuming meat, are necessary for the growth and development of the human body. Usually, humans consume the meat from poultry, mid-sized animals like goats and pigs, and large-sized animals like cows, donkeys, camels, and so on. These are the main sources of meat for daily consumption and are commercially produced on farms. Wild, as opposed to commercially produced meat, may have value and benefits for human health; however, more research is needed in this area.

Preservation of meat is important and typically accomplished with refrigeration. However, refrigeration requires large amounts of energy. New advances in food storage along with traditional techniques may lead to the development of low-energy solutions that preserve the nutritional value and taste of meat. In other words, it is necessary to produce and preserve meat at low cost at both individual and commercial levels so that all people have access to affordable meat with high nutritional value (i.e., contains essential amino acids). There is a great disparity in meat access and consumption. Wealthy people have been accused of overconsuming meat, whereas less wealthy people may not even have access to meat.

The ultimate goal of consuming meat is good health. There are many meat-borne diseases due to unhealthy handling and consumption. Recent evidence from large prospective US and European cohort studies and from meta-analyses of epidemiological studies indicates that the long-term consumption of increasing amounts of red meat and particularly processed meat is associated with an increased risk of total mortality, cardiovascular disease, colorectal cancer, and type 2 diabetes in both men and women. The association persists after the inclusion of known confounding factors such as age, race, body mass index (BMI), history, smoking, blood pressure, lipids, physical activity, and multiple nutritional parameters in multivariate analysis. The association has not always been noted with red meat, and it has been absent with white meat. There is evidence of several mechanisms for the observed adverse effects that might be involved; however, their individual role is not defined at present. It is concluded that recommendations for the consumption of unprocessed red meat and particularly of processed red meat should be more restrictive than existing recommendations. Restrictive recommendations should not be applied to subjects older than 70 years of age, as the studies quoted herein did not examine this age group and the inclusion of sufficient protein supply (e.g., in the form of meat) is particularly important in the elderly. This is a challenge for public health.

To promote people's health, existing and ongoing research is not sufficient. Clear policy interventions are also needed. Poultry and livestock contribute greatly to the GDP of countries. Globally, livestock production currently accounts for some 40 percent of the gross value of agricultural production. In industrial countries, this share is more than half. In developing countries, where it accounts for one-third, its share is rising quickly; livestock production is increasing rapidly because of growth in population and incomes and changes in lifestyles and dietary habits. Thus, it is necessary to formulate policies on meat production as well as consumption to balance public health and the economy.

This book covers a wide array of topics related to meat and nutrition. Chapters present recent research on the production, consumption, and handling of meat. They also explore the nutrient value of meat that is not commercially produced. This book is important for academicians, researchers, policy makers, and entrepreneurs who want to invest in poultry and livestock.

Dr. Chhabi Lal Ranabhat, Ph.D.
Health Service and Policy,
Department of Public Health,
Manmohan Memorial Institute of Health Science,
Kathmandu, Nepal

Global Center for Research and Development,
Kathmandu, Nepal

Pai Chai University,
Daejeon, South Korea

Section 1

Production and Consumptions Pattern of Meat

Chapter 1

Inequality: The Dangers of Meat Haves and Have-Nots in a Nicotinamide-Adenine-Dinucleotide World

Adrian C. Williams and Lisa J. Hill

Abstract

Our evolution and recent history can be seen as a "World Hunt" for meat as part of an omnivorous diet. Meat contains key micronutrients namely Nicotinamide (vitamin B3) and methyl-donors with deficits causing pellagra, an archetypal disease of poverty. Inequality is a leading ultimate risk factor invoked in the aetiology of common diseases let alone threats from climate change and pandemic triggered catastrophes. We hypothesize that the origin of inequality was our evolutionary and nutritional move from equal to unequal sharing of the meat supply some 10–20 thousand years ago. High meat intake may have bioengineered powerful ruling classes and lower intake the proletariat with higher fertility, but inferior (brain) health. A fairer quantity of a safer meat intake in future should moderate global variances of fertility, height, health, and prosperity. Death rates of acute infections including emergent zoonoses (such as COVID-19) and chronic infections (such as TB) should fall as might the incidence of some diseases of affluence. Meat justice by improving human capital could make redundant superficial markers, such as skin colour, used to discriminate against peoples and heal a divided world.

Keywords: Disease Transitions, Demographic transitions, Anthropocene, Nicotinamide, COVID-19, ACE2 receptor, Tryptophan, Multiple sclerosis, Tuberculosis

1. Introduction

Prelapsarian human nature was egalitarian sharing animal products that are the main sources of nicotinamide, tryptophan and methyl-donors. The origin of inequality was in the Mesolithic with unequal sharing of meat creating phenotypic variety in a genetically homogeneous population (genomes were later modified by nutrition and infection [1]). A high meat intake allowed for a ruling intellectual class and a lower intake worker class with higher fertility but poorer health. Meat intake currently manages hundredfold variances within a global annual 300 million metric tonnes (was 7 million in 1960 and could rise another 75% by 2050). Meat inequality is high and for billions their slice of the "meat-loaf" is wafer-thin undoubtedly affecting their well-being. Wells, 2016, threw down the gauntlet: "If we cannot define the link between nutrition and power we will never gain the power to resolve global malnutrition and its numerous costs" [2].

2. Extreme meat inequality: the forgotten case of pellagra

Indeed inequality is generally held to be the pernicious culprit responsible for many medical and social ills faced by food-insecure billions that can lead to trade-offs between survival with high fertility but poorer health and shorter lives [3–8]. As defined by Bellamy (1897) the basis of equality is when "*...there are no more a-hungered*".

An iconic example of a nutritional trap is when a low meat intake risks the degenerative condition pellagra whose sufferers, with inferior cognitive and social intelligence, were ostracised as the "Butterfly caste", and contracted infections such as tuberculosis (TB) also closely linked with poverty [9, 10]. Terms used to stigmatize, shame, blame and pillory pellagrins are still in common usage today to keep the poor in their place. Worse was the call for forced sterilization based on eugenic and racist policies building on the "myth of the lazy native". Yet there turned out to be a biological and trans-generational explanation for this man-made layer of destitution preventable by public health means.

3. A desire for meat

Nutritional traps drive a "flight to quality", as noted by Ernst Engel in the 19th Century [11]. As the price of bread falls or when incomes rise people spend less on starches but more on meat up to a point. This gastronomic desire extends to cannibalism documented in the Magdalen (30,000 years ago) as funerary defleshing and later ritualised by states short of meat in Central America or, as infanticide or witch-hunting [12, 13]. Cannibalism has proponents for a "materialist" theory and the need for protein but it is also a symbol of "savagery" giving many an excuse for racism, slavery and "civilising" colonialism [14]. In retaliation cattle-based original capitalism and its descendant expropriations of land and nature has been convincingly called "cannibal capitalism".

Rich Americans eat more than their body weight in meat every year whilst many in the "Global South" are on negligible amounts. Developed countries are not immune as their poor, often children and minorities, fall below "Eat Well Plates" as witnessed by the rise of food banks and the recognition of place based food deserts where good food is unavailable. This also creates (obesogenic)socio-ecological environments that argue against neoliberal paternalistic views on the incompetent poor having "mismanaged lives" that need to be disciplined or shamed, stereotyped as "chavs" and stigmatized by "body fascism" or politicized by neoliberals as "deplorables" - as were pellagrins in their pathological NAD-deficient food-scape in "Foucaultian" fields of lost-power and little choice. Geographical meat transitions are still occurring though not everywhere: in 1962 the average Chinese was eating 4 kg pa but now that figure is 60 kg pa and rising fast towards the American average of 120 kg pa. Ten calories of animal feed produce 1 calorie of meat and need enormous quantities of water, oil, fertilizers, pesticides, and antibiotics let alone consumption risking dangers from food poisoning and zoonoses with human and economic costs [15, 16]. Given all that, and given animal rights abuses and that meat producers are high contributors to global greenhouse emissions, one would hope that there is a sound biological demand rather than a higher supply on the market for "showing off" .

4. Demography and subsistence are key considerations

Modes of subsistence and demography are the place to start a quest for the source of inequality [17, 18]. Malthus noted that poor parishioners reliant on cereals had high rates of baptisms relative to burials sparking concern that their high fertility led to cycles of deprivation [19, 20]. He commented on the sparse numbers of the more carnivorous hunter-gatherers and that population densities increased exponentially with cereal based agriculture. Conversely Boserup suggested population pressure increased agricultural innovation to cope and De Castro's "Geography of Hunger" (1952) pointed out reverse causation was at play in that global epidemiological and experimental data suggested that a degree of malnutrition increases fertility and quoted Doubleday's "True Law of Population" (1853) on high meat intake decreasing fertility.

Fertility may have a "U" shaped relationship with meat intake. Low nicotinamide in diet leads to its synthesis "in house" from the degradation of tryptophan. This pathway is an "immune tolerance" mechanism that can welcome foreign antigens such as the foetus or symbionts, but risks dysbiotic and acute infections - and may switch to immune intolerance as the nicotinamide dose increases [14, 15]. Teleologically this allows "baby booms" as diet improves when emerging from famines and for slight changes in fertility that compounded over generations alters trajectories from extinction to strong growth and for shifts toward quality over quantity of offspring [21, 22]. Disease inequality could derive from subpar meat intake and nicotinamide related biochemical and epigenetic mechanisms to affect "human capital" with other life-history trade-offs and dietary mismatches over lifetimes then forming the developmental origins of adult disease (DOHaD) and late-life and transgenerational inequality [23, 24]. Current demographic and disease correlations with factors, such as education, may be hiding a "lurking" variable of food, particularly meat, resource; this systemic dietary inequality was not present in our "deep" history [25, 26].

5. Meat and brains: "Planet of the Apes"

Primordial pecking orders with dominant alpha males or females were more over access to mates. At the time of the "Great Divorce" *Homo* increased meat intake,

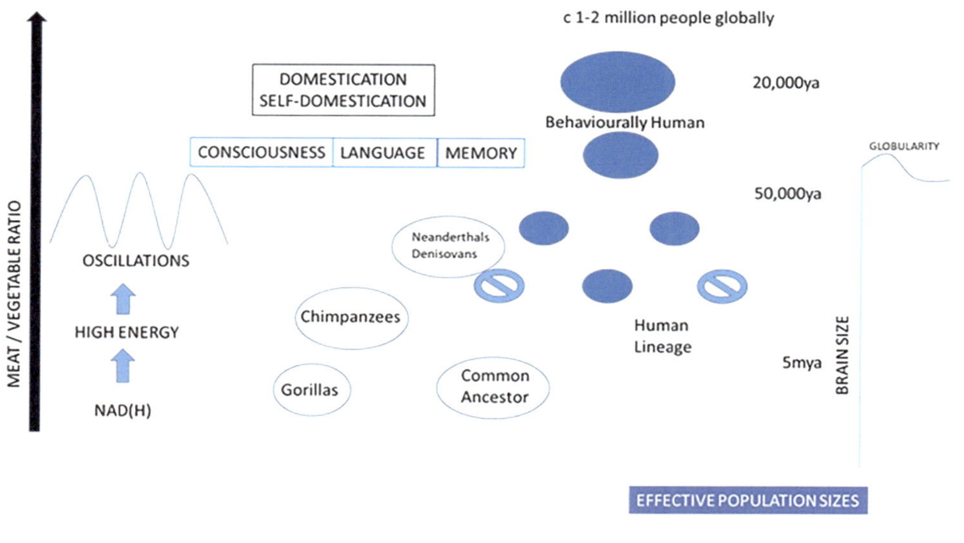

Figure 1.
Meat and Nicotinamide dosage steadily increased during our evolution up until the time that we became behaviourally modern. Human brain size increased and got more globular with Broca's and pre-frontal and parietal areas becoming prominent and better connected using newfound neurotransmitter and neuroendocrine facilities. However fertility and population sizes were low, with several extinctions. The advent of a more plant based, and lower nicotinamide dosage, diet led to populations expanding but brain and body size got smaller and infectious diseases emerged.

sourced on the savannah, became reproductively isolated ("kissing cousins" on forest edges excepted) and at a fork in the road speciated (**Figure 1**) [27, 28].

6. Food and fortune

Trans-continental food quests with the prosocial and technological skills for hunting catalysed the NAD(H) based energy rise required for high general intelligence in positive feed-back loops [29, 30] (**Figure 2**). Hunting parties crossed the globe extirpating animal, bird, fish, or sea-mammal species in their wake.

Homo sapiens and Neanderthals independently evolved large brains on high meat diets but both species were "thin on the ground" with populations that "tottered" with local extinctions and population bottle-necks that led to the exponential expansion and cultural flowering of one but the simultaneous extinction of the other [31, 32]. *Homo sapiens* honed in on the difficult to digest and toxic plant foods detoxified by cooking and xenobiotic enzymes in a cultural and genetic co-evolutionary approach [33, 34]. This move down the food chain along with pro-fertility cultural innovations, exemplified by cosmetic ornamentation and seductive figurines, perhaps rescued us from extinction [35, 36].

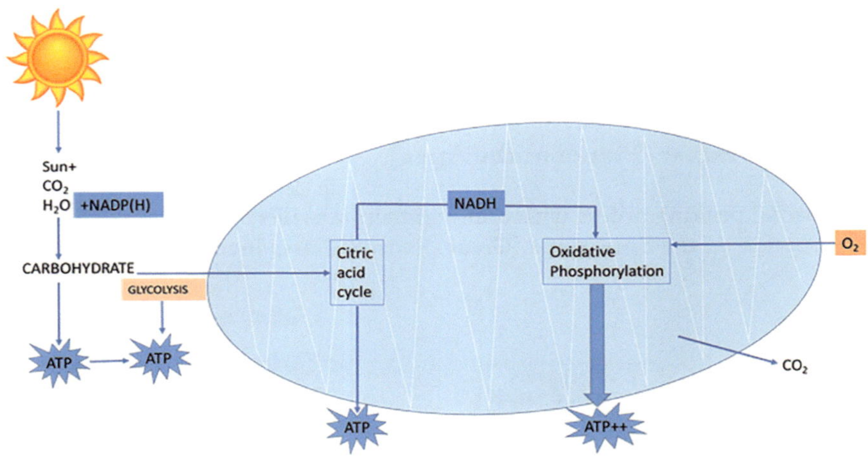

Figure 2.
NAD is the crucial carrier for our high energy Hydrogen based needs for optimal brain function in a "NAD World".

7. At human evolution's heart was meat-centred equality

Hunter-gatherer social norms were egalitarian sharing meat with kin and non-kin, at least within the reproductive in-group. Land was then a shared "commons". Social animals fight for the spoils even when they are by-standers - so this was our "social leap". Leaders only existed for time limited tasks. "Stag Hunt" and "Ultimatum" games demonstrate a residual sense of fair play in contrast to the misanthropic "Homo economicus" depicted in the "Tragedy of the Commons" [37, 38]. This redistributive system created the most long lived economy in our history with ample leisure time and was the dietary evolutionary environment to which we adapted [39]. Adaptations have occurred since (such as lactase persistence) but a mismatch with this "Palaeolithic" diet may still be relevant to modern day illnesses - particularly for the poor or the post-reproductive who are of an age when selective pressures to adapt are attenuated implying that their metabolism, in particular, would perform better on the long-abandoned ancestral diet [40, 41].

8. A more variable subsistence package developed

Horticulture emerged in the Mesolithic in marshlands and uplands. Communal village "nests" allowed storage, helped by pottery, and pans for vegetable and meat stews [42, 43] and veneration of fertility and diet - later examples were Ceres, Maize, and Bull cults [44]. A sexual selection process included language, dance, laughter and cooking domesticated and "civilized" us encouraging our reproduction and controlling the reproduction of domesticates [45].

9. The "great disequalization": outer walls inner castes

There was a lag of some 5000 years between gardening and Neolithic agriculture and aquaculture that started in arid zones between rivers suitable for irrigation or flood-retreat alluvial zones. Another long gap exists before city and national walls. Walls kept out pastoralist egalitarian barbarians and their meat surpluses traded or raided for grain - and kept in a populace with their cereal surpluses that could be taxed by rulers [46]. Cities record social stratification with kings, priests and military elites feasting on quality foods and waging wars over meat resources. Nobles were taller and healthier and better educated as a "cognitive class" not unlike our well-fed "meritocracies" [47, 48]. This disequalization event perhaps started earlier in a mosaic such as in the sedentary Nafutian culture but wherever it occurred a relative shortage of meat fits the facts well: inequality even developed in non-agricultural communities who needed technological advances such as ocean-going canoes or horses to hunt new sources of meat as it ran out [49].

Much has been made of class differentiation in Eurasia being more over the quality of food but over the quantity of food in Africa however if meat is the crucial factor, and manners, spices and sensuality more superficial, this paradox disappears as meat was more of a luxury in equatorial Africa [45, 50]. The importance of meat is shown by cattle as capital with transfers in "bride-wealth" dowries and as a universal central-dish in feasts [51]. Crucial determinants of inequality were ownership of land and livestock that could be inherited with Gini coefficients as low as .25 for foraging hunter-gatherers compared to .5 amongst agriculturalists.

10. Stocks and trade: an overdue tribute to "Barbarians"

Savvy pastoralists at independent sites developed dairy that as a source of nicotinamide riboside could explain the convergent genetic evolution of lactose tolerance and the cultural evolution of fermented yogurts and cheeses [52, 53]. Steppe peoples and their ideas spread across Europe around 2500 BCE, replacing or amalgamating with agriculturalists as did later mounted pastoralists [54, 55]. The fall of the Roman Empire on a diet of "bread and circuses" and many pandemics allowed Germanic pastoralists with their pedigreed animal husbandry to overwhelm a cereal dependant system (with its "agri deserti") and Roman deserters [56].

11. See-saw cerealization: meet thy maker and breaker

A Green revolution around 1000 AD with unification of African and Asian crops now with rotations and multiple planting seasons during a warm medieval period allowed further "Cerealization and Calorie-ization". The social gulf between meat-eaters and grain-eaters was a cultural fact of life with social penalties for transgressors [57, 58]. Populations boomed then busted with the Black Death [59] then recovered slowly on the higher meat diet available to the survivors whose better human capital may explain the rise of Europe.

12. Old and new worlds: all things (NAD) were not equal

American megafauna, as in Australia, had unlike the "Old World", no prior experience of resisting human predators leading to their easy extinction as the hunters arrived 10–15 thousand years ago. The New World thus had less animals in general and were unlucky with their limited choice of domesticates, given no sheep, goats or cattle. Comparison between Old World social structures and the New World shows that the latter were the less stratified with less inherited wealth [60]. Old Babylonia yields a Gini of .40 whereas near contemporaneous Teotihuacan scores a low Gini of .12. Similar observations were made in China with its low level stratification and pigs but no draft animal's supports availability of "food on the hoof" as the driver rather than animal labour. This all suggests a "U" shaped curve with high and low meat intakes favouring egalitarianism and collectivism but somewhat constrained meat supplies leading to stratification (later in North America an abundant meat supply was an explanation given for the lack of socialism and high stratification "on the shoals of roast beef and apple pie" [61]).

The Columbian exchange exported maize and tubers east in a non-uniform fashion, driving local population explosions. In exchange ungulates were introduced to the New World. Breeding rates were extraordinarily high so much so that ecological damage was caused by often feral "plagues of sheep" (that compares with "plagues of corn" in Europe). 17th C Spanish and Portuguese ranchers maintained herds of 7–10 million animals producing a surfeit of veal in industrial scale pastoralism [62, 63]. However introduced zoonotic diseases, such as smallpox, decimated local populations immunologically weakened by their low meat/high cereal diet as much as by lack of "herd resistance".

Observers noted that as meat intake increased Native Americans health improved and they became, they thought by Galenic "humoralism", more Spanish, partially reversing concerns about racial decline with inter-marriages but still creating new castes with the poorest Amerindians displaced to reservations unable to hunt [64]. One astute writer (1596) presciently noted that "meat generates superfluous humours so they now sneeze as we do" suggesting an early switch from infectious to allergic disease - a phenomenon repeated in the late 19th century as meat intake recovered from an earlier fall in Europe as we discuss later [65, 66].

Maize went east as an important part of the Columbian exchange but of all the cereal staples it has the lowest concentration of tryptophan and nicotinamide so much so that there was an evolutionary drive to cook in a (female)labour intensive process with alkali producing "nixtazmel" in Mesoamerica; but this culture or even mixed planting and eating with beans was not exported east putting those in the east at a higher risk of pellagra – despite this maize was popular as it adapts to variable altitudes and water supplies with high yields unlike wheat or rice [67, 68].

By contrast with successful pastoralists then nowadays many herders are poor. This reflects changes in the meat market with more advanced societies distancing themselves from zoonotic risks by industrializing meat production. Pastoralism *per se* is no advantage unless it allows the owners a higher income or access to their own animal source foods free of contamination [69].

13. Meat elites: NAD "us and them" co-operations and conflicts

We argue that a sliding rule of meat intake benefits states as well as classes by engineering upper "expert" classes with high longevity (adding to their crystallised intelligence) to the lower classes with their "essential" but often poorly paid and dangerous front-line jobs, but higher fertility. As Henry George said in 1879 *"This association of poverty with progress is the great enigma of our times; not to answer is to be destroyed."*

At a more macro- level a latitudinal gradient in food-getting technology to catch prey in the more animal dependant climes exists and once weaponised fuelled northerner's fire-power as perhaps did their more individualistic culture [70]. Luminaries such as McNeill and Maddison mention transatlantic meat flows alongside technological nous in their expositions on the rise of Europe [71, 72]. Colonialism and World Wars aimed to ensure enough pastureland for the winners and at the same time cutting off the colonies or enemies food supply inflicting developmental and epigenetic scars on the losers, as documented in the Danish "Hongerwinter" of 1944 [73, 74].

14. Skin colour and nicotinamide

Variation in human skin pigmentation, whether from genetic polymorphisms or tanning, is the most important physical trait used to instantly categorize human groups and individuals [75, 76]. Pale skin has the adaptive advantage in low UV environments for vitamin D production. Darker skin protects against the rash of pellagra and the closer to the equator the more populations were at risk as the meat/vegetable ratio falls compared with temperate and polar climes. Resistance to the rash is good short-term but as it serves as an early warning to (self-)treat before more serious and harder to spot effects on cognition it may be disadvantageous at a population level and opens a door for discrimination.

The idea of intellectually and morally inferior races based on complexion (that otherwise seems absurd), accelerated with the scramble for Africa and Atlantic with slave-owners conveniently believing whites and blacks were different species - views that others did their best to dispel *"God hath made of one blood all nations of men"*. Links with low meat intake go back to Saharan trades with captives turned to slaves from civil wars usually over the meat supply as equatorial pastoralism is harder. Local ungulates resisted domestication and are threatened by large carnivores and year round transmission of vector-borne diseases in the vast tsetse fly belt - and by rapid proliferation of pathogens in food in the heat.

Many believed they were sold for cannibalism but in fact died in droves in the sugar plantations of the Caribbean; in the Americas they were fed somewhat better such that fertility rates allowed for generations to be born in slavery - but were not so well fed as to avoid pellagra particularly after emancipation and neoslavery [77–81]. Policies directed at indigenous and imported peoples were early assimilation or attempted annihilation if expropriating hunting lands. "Buffalo Bills" executed their bison but "last drop of blood" segregationist policies allowed reproduction if more after labour - either policy conspired to deliver an inferior diet for many [82]. In contrast to the Comanche and their colleagues, cattle now fenced in by barbed wire on ranches and ranges and protected in a "6-shooter colt and cowboy empire" created a beef and red meat republic. Industrialized meat processing, as in Chicago, followed with an international capital market aided by steam railroads and ships with refrigeration.

Confederate cotton states that housed pellagra were in the forefront of supremacist "White privilege" "Klansman" and "America First" thinking. The common interests of this multi-colored underclass were muted by racial tensions encouraged by white elites to divide and rule the workers and were even written into national and state constitutions and (Jim Crow) laws. W.E.B. Du Bois writing after the American Civil War referred to a divisive dignity with being white seen as a substitute for inclusive economic policies that could have improved diet for all assembly-line and other workers: degradation of black labour being seen as more important than uplift of white labour. Even the 1890s Farmer Alliance bottom-up populist movements were weakened by segregation and racism and undermined later "Wars on Poverty".

Others were not immune as poor Italians, Irish and Gypsies or even alcoholics in degenerate "drinking classes", also prone to pellagra, are often considered inferior races [83]. Genocidal thinking against others, such as Jews or the Tutsi tribe, may be because they were thought superior but these are historical exceptions as are those examples of collectivist and communist anti-middle class agendas, such as in China, Russia or the Cambodian Khmer Rouge. Most of the rest are subject to well fed "White Anglo Saxon Protestant (WASPs)" and Western, Educated, Industrialized, Rich and Democratic (WEIRD) people being in charge though this in reality may allow for the mediocre to flourish. Diet and type of agriculture when contemporaneously studied across America or across countries affects cultural norms from "tightness" to a "looseness" that supports a more individualistic and entrepreneurial society with extreme wealth inequality – "tightness" maps closely to former pellagra states or cereal based cultures and collectivism with a high incidence of chronic infections and other signs of poor development [84, 85].

15. Beyond the pale: pellagra and the undeserving poor

The "undeserving poor" whether amongst white skinned "Hillbilly" rural classes in America or in England (originally noted by Cobbett in 1872) were prominent sufferers from pellagra and like poor blacks attracted the attention of eugenicists and social Darwinism although, to be fair, more positive "social hygiene" ideas targeted diet and education [86]. Developmental impairments may have spawned the "sciences" of phrenology, physiognomy and craniometrics that helped create myths about black racial groups having deficits in brain capacity.

Push-back has occurred with peasants, slave ("Black Spartacus"), and many indigenous people's revolts although poor diet may weaken resistance. Pellagrins had specialist trade unions and newspapers "Il Pellagrasso" and, driven by "Pellagraphobia", "Pellagrasorium" hospitals. School meals welfare programs have a surprising history for example in being promoted by the activist Black Panthers despite attracting heavy opposition from the FBI who perhaps realized those at the knife-edge had got to the heart of the matter of connecting diet to power and the political economy [87]. The rise of the middle classes and enlightenment thinking on food and the first restaurants insisted on regimens elaborating on meat and 2 vegetable based diet [88, 89]. Frustration such as by the 20th C solidarity movement in Poland was driven by annoyance at queueing, often unsuccessfully, for meat that eventually freed them and others of the communist yoke [90].

16. Poor immigrants emigrating for meat

"Out of Africa" hunting parties from around 70,000 years ago (and earlier for our hominid ancestors), was driven by the need for meat. Later meat food-ways in the age of migration and the "hungering for America" came from groups known to be pellagra prone such as the Irish, Italians and Mexicans. Once arrived, they ate like the aristocrats they had left behind. Similarly the African-American northern "great migration" around 1879 of some 6 million freed "Exodusters" were fleeing from the pellagra-prone southern states. The initial poor state of all such immigrants, that included smallpox outbreaks in slums, contributed to xenophobic discrimination as did their high fertility setting off worries about degeneration and displacement of the local whites [91, 92].

17. Gender, religion and nicotinamide

This worry overlaps with gender inequality that explores a similarly dark history. Female sex, like colour, compounds risk factors for pellagra with men, the "bread-winner bringing home the bacon" and also the "carver" controlling and rationing the meat amongst family members being given priority over women. This long standing dietary disadvantage and lost privilege over meat rations may have increased fertility but could have spawned much male entitlement including to sex (sometimes traded for meat) [93]. High fertility, as already mentioned attracts criticism as "Welfare Queens" and the attention of eugenicists, family planners, and as a part of "Great replacement theory" these worries intersect with antipathy to rival religions that promote reproduction and rely little on converts.

18. Occam's razor: real bias is against the less educated

Intersectional and multiplicative effects of these injustices and many exceptions from superficial markers, that may reflect the cultural schisms and "identity politics" of the day, is compatible with a common more material and tangible cause in diet. Indeed the politics of recognition may at times be at odds with the political and human need for redistribution. Diet induced poor cognition that, if unrecognized, neither allows for equality of opportunity or for society to show solidarity with those who do not rise (even though essential workers), leading to their segregation or even incarceration [94–96]. Data suggests that the college educated "meritocracy" (usually well-fed), have more bias against the less-educated than they do against any other dis-favored group as a "tyranny of merit" . This is even true of America's black upper class that originated in freed slaves, or because they worked inside the master's house, that had a better diet than field slaves and more access to educational material. Dietary differences could explain disparities between communities given that success differs between black Caribbean's and black Africans with both performing better than poor whites and neither better than rich Asians or rich Whites. Lower IQ, often in the "Imbecile" ran were core features of "pellagra sine pellagra" who frequently failed the very basic tests required to join the military. A good diet was important to the evolution of "WEIRD" people [97]. The net track record of such intellectuals realizing they are part of a "meat elite", rather than having a superior genetic or racial endowment, or in sticking up for the poor or racial groups or falsely believing in an overriding role for artificial selection is a classic "trahison des clercs" [98].

Dietary head starts also define Diamond's milestone hypothesis on global faunal inequality with "lucky latitudes" for farming at the onset of the Anthropocene.

19. Meat inequality: the climate link

The origin of the climatically benign Holocene heralded the "Anthropocene" that consists of a series of horticultural and agricultural developments - some even call it the "Plantation-ocene" [99, 100]. The Anthropocene influenced climate by deforestation and terraforming affecting CO_2 and methane emissions from rice production and animal domesticates keeping the benign Holocene climate rolling [101–103]. These arguably reversed temporarily after the pandemics of the Columbian collision - as the 1610 "Orbis spike" – and a "Little Ice Age". An unhomogenised intercontinental meat supply and green agricultural advances has ever since driven population explosions of both domesticates and ourselves. Alongside the advent of fossil fuels and artificial fertilisers these have conspired to become major contributors to climate change with further inequality in ruptured "Sacrifice Zones" characterized by low to negligible meat intake variances that make for both a "Meat-obscene" and a "Planet under Pressure."

20. Farewell to alms – one for all and all for one

Dietary variances may allow some wanted diversity and plurality but meat became the origin of inequality however this was against strong resistance as reflected in a fitful history over the right for a balanced diet that we will now summarise [104]. As has been said *"The arc of the moral universe is long but it bends towards justice."*

Aristotle first proposed that government provide good nutrition by means tested communal meals and that private land could be used by people in need so that all could flourish. Utopian thinking pleading for public help for paupers such as by 4thC Saint Ambrose –*"the earth has been created in common for all, rich and poor"* – and the 13thC Thomas Aquinas and 16thC Juan Vives and Thomas More argued that stealing if hungry was not a criminal act with the latter in his Utopia (1516) first suggesting a Universal Basic Income. Later John Locke (1689) a strong supporter of the state protecting the sanctity of private property rights excluded cases of "pressing Wants" where stealing if hungry could be justified - *"God hath not left one Man so to the Mercy of another, that he may starve him if he please"*. Thomas Paine (in 1796 irritated by a bishop preaching *"God made rich and poor"*) argued for redistribution *"not bounty but justice"*- not with scraps, crumbs or handouts but compensation for lost farmland to *"buy a cow and to cultivate a few acres"*. Howlett however insightfully felt that opposition came from a gravitational pull to increase fertility and to create a paid male and unpaid female labourer class to provide social reproduction [105].

There was further intellectual support in early "socialist" and (French and American) revolutionary thinking of provision as a right not as charity. Thomas Spence's pamphlet ("The Rights of Infants" 1797) and Charles Fourier are good examples –*"If the civilised order deprives man of hunting, the class that took the land owes to the frustrated class abundant subsistence"*. Von Humboldt with like-minded agrarians including Goethe and Jefferson and Madison in the infant USA understood the effects of colonialism and deforestation and the need for less parasitic approaches to nature bucking the biblical "*dominion over all the earth and every creeping thing*". Many empires encountered local resistance and insurgencies such as the Indian Mutiny of 1857 with early dissent from universalist thinkers who eschewed biological racism and believed all men to be equal such as Burke, Bentham, Smith and Diderot (1780) were concerned about European explorers, pioneers, and colonialist unjust and attitudes *"instead of recognising this man as a brother, you see him as a slave"*. This enlightened attitude later lost out to civilising missions of "backward societies" and the frontier spirit, supported by Mill and de Tocqueville, and racial ideas of white superiority mitigated but not solved by Wilberforce and the anti-slavery movement or the American civil war.

21. Enclosures, empires and the "third world"

Oppositions to underhand removals and expropriations of common pastureland from serfs are recorded. Resistance included the Magna Carta (particularly the Charter of the Forest (1217) that talks about "common herbage") and the 17th Century leveller movement and opposition to the notorious Black Act (1723) [106, 107]. Poachers and commoners even blackened their faces to disguise their identity and to show solidarity with slaves. Nevertheless Arcadian grasslands got eroded by the "enclosure" movement and punitive laws for poaching and the birth of "*Enemy of Nature*" capitalism with its exemplar lack of recycling manure as natural nutrients back to the soil and "metabolic rifts" as first proposed by Marx. Enclosure of pastureland is also associated with the concept of "social closure" when scarce resources only get shared with those of the same class such as certain clothing and education – and the rich monopolising a gourmet taste for meat [64].

Dietary ideals sank into oblivion with imperial grabs of land creating "new Europe's" with "cash crops and stocks", mining of bones from Napoleonic battlefields and importing guano for fertiliser, and the "triangular" slave trade. Governments and companies employed armed forces to crush uprisings with "scorched earth" campaigns leading to famines and genocides creating the third world by kyboshing local development and introducing pellagra-genic maize [108, 109] (**Figure 3**). Imperial interlopers farmed then imported cattle or taxed locals providing themselves with a 'free lunch' resulting in "slow violence", "long dyings", "zones of abandonments", "necropolitics" and "tristes tropiques" and "Victorian holocausts" with both ruins and ruination [110]. Other plunders and blunders include the ugly histories of the Irish famine, the Scottish Clearances, the Soviet war on the Kulaks, the US "dust-bowl" and the Chinese Cultural Revolution. Colonial near starvation led to debilitating phenotypic adaptations (in survivors) often acquired in childhood in "metabolic" ghettos, such as by Native Americans and Aboriginal peoples thrown off their hunting lands; or later as in the legacy in the Caribbean of a low meat/high sugar diet followed by a western diet triggering the "double burden" pandemic of metabolic ("amputation capitals") and cancerous syndromes [111]. Slave trade reparations were not given to the slaves or to their epigenetically affected descendants however there is some history of trying to help the poor locally [112].

22. From poor laws to meat rations

Elizabethan poor laws were a reaction to the dissolution of the monasteries and a resurgence of "Royal Forests" that reduced common pastureland. The 1834 poor law with workhouses and means testing legitimized the concept of the undeserving poor and resulted in Edwardian slum-dwellers being no better off than the later

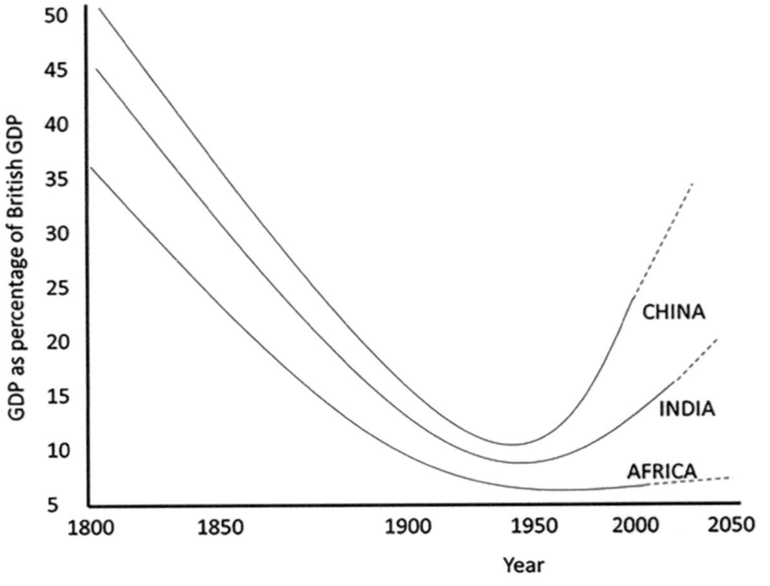

Figure 3.
GDP falls as a % of British GDP became extreme in colonial times. Low meat diets in China, India and Africa compared to Europe and North America created the "third world". This dietary inequity is unravelling in places with the "tiger economies" undergoing "meat transitions" developing the fastest.

starving victims of Somalia or Rwanda. Poor diet came to the fore when the state of recruits to the Crimean and Boer wars affected the country's defenses with hunger marches adding to the pressure.

Initiatives such as a broader diet in WW2 rations and school milk and meals improved health and infant mortality as did "cradle to grave" welfare states. Lessons on the primacy of diet still got forgotten and never rolled out internationally despite experimental evidence that poor diet influenced individual, class, tribal and national success [113].

More evidence on diet comes from the Indian caste system as the lowest untouchable class (Dalits) in a "metabolic ghetto" were short and unhealthy on rice and vegetables compared with Brahmins (who ate nicotinamide rich buffalo milk, yogurt and butter) and other castes on wheat and meat. In Kenya the meat and blood eating Masai were taller and healthier than the vegetarian Kikuyu tribes, who suffered greatly from TB. Specific mention was made of the near impossibility of modernising in the Caribbean on a plantain diet yet botanical benevolence, such as introducing sago plants and breadfruit, was commoner than promoting meat perhaps as the immediate pressure usually seemed to be about bread.

"Flour wars" have triggered the downfall of empires and aristocracies such as in 18th C France and early 20th C Russia and along with the British experiences in Ireland and Bengal and the recent bread riots in the Arab Spring uprising suggest that the food supply chain is an iceberg underlying stable societies and financial markets. Governments and commerce should aim higher than avoiding caloric starvation [114]. Indeed WW2 rationing was thought to have made class war obsolete with a nutritional egalitarianism, that covered meat and milk, and led to a 30 year upswing in equality lasting long after the normal levelling effect of the exigencies of war [115, 116]. This temporary upswing included "sharing the prize" with black southerners in America helped by the civil rights revolution that had not happened with the 1930's New Deal that was, despite some good aspects, racialized on housing and jobs and therefore the income to buy meat [117, 118].

23. Tiger economies – a unified field and food theory

The age of Industrialization increased the gap between the North Atlantic states and the rest of the world: the former had high meat intakes with the "laggards" being cereal dependent. Japan overcame Buddhist piety that proscribed consumption of four legged animals and both imported beef and altered their class system. Later "Tiger" economies built arcs of food security less hooked on subsidised cereals and more generous on the more elastic need for meat. They realized, or were advised, to "use it (their land) or lose it" risking become "banana" republics. The lesson of the 19th C Ireland "meat republic" is apposite as the Irish landowners exported cattle to the UK whilst their own cottager population boomed on a poor potato diet until blight led to widespread starvation and emigration [119, 120].

China followed suit, after disastrous collectivist experiments when some 45 million people starved, and massively increased meat consumption surging to the forefront. India have followed but with lower increases in meat consumption (and lower growth), as has Latin America but not sub-Saharan Africa. Cuba managed with modest increases in meat consumption to demonstrate beneficial effects on measures of health and happiness [16, 121]. Such countries achieved modernity with no significant aid that usually came as subsidised cereals or the "Green Revolution" as in much of Africa [122]. Cereals and sugars along with apartheid thinking of Africans being inherently poor unscientific farmers in "cattle complexes" considered as wealth not food in a "malnutrition syndrome" (whilst valuable food is exported) to create a vicious cycle leading to "starving on a full stomach" and micronutrient deficiency, including B3/Nicotinamide and pellagra outbreaks particularly amongst refugees from war. The paradox here being that Africa has plenty of sun and enormous land-banks but their agricultural methods and utensils would have been familiar at the time of Christ creating crop yield chasms with knock-on effects for animal fodder and meat intake.

Tables have been turned in that food exporters are now in the rich world that subsidises its farmers with the poorest countries off-shoring even grain staples risking international food spikes. "World-making" needs more international effort than expecting self-determination to help with diet and could be seen as a practical reparation [123]. After all, the development of a European core was given priority over colonial settlers raising cattle for sale at the centre at prices that excluded the peripheral colony and allowed the industrial "take-off" [124, 125]. The rise of Anglo-American hegemony and the current convergence in a predominantly Asian drama that also correlates with meat intake could be enacted everywhere to help demographic and disease transitions.

24. Levelling playing fields

If looking backward to imperial violations provides no traction risk of pandemics and wars may be the better bargaining tool as poor countries are not, after all, stationed on Mars [126]. The history of disease and demographic transitions when the West was just as poor is instructive as progress correlated then to an increased meat and milk supply and the colonial "klepto-parasitic" meat-trade [125, 127–129]. As Walter Rodney tellingly said in his 1972 book on how Europe underdeveloped Africa *"Pellagra was unknown in South Africa till about 1914"*.

Many have commented on the importance of meat and skimmed milk on health in particular the incidence of TB – and as a cure for Kwashiorkor and is the basis of many school milk and meals programmes. These early 20th C programmes often driven by fear of TB were sometimes reversed such as in 1950's south Africa for African but not European children as they were "white man's food!" [111, 130].

25. Beefed up: Au Revoir "Old Friends" and Plagues

It is difficult to overestimate the pervasive importance of TB the "White Death" in the 19thC that mysteriously vanished (as did other infections) first in the wealthy as Disraeli pointed out *"Two nations: as if inhabitants of different planets formed by a different breeding and fed by a different food – the rich and the poor"*. At this time food imports (the UK at this point accounted for 80% of the trans-equatorial meat trade) were aided by lower shipping costs, trains and salting then refrigeration [118, 131, 132] (**Figure 4**). Better breeding helped as did the rise in the use of poultry. The case for nicotinamide intake being causal has been that TB excretes and is inhibited by nicotinic acid with many antibiotics being analogues and that TB incidence always rises on a poor meat diet [133, 134]. TB's toxin, an NAD glycohydrolase, depletes the macrophage of NAD on a cell-death pathway that enables replication and dissemination. Over 300 like toxins are responsible for other pandemics [135, 136] so optimal NAD levels offers "broad spectrum" protection against many organisms that is lost if diet then deteriorates.

26. Inflection: inflammatory disease in affluent geographies

As TB died down a promiscuous range of auto-immune, inflammatory, and mind altering "Diseases of Modern Civilisations" took-off alongside infertility, first in the upper classes who eat more meat [137, 138]. A less plant based diet affects fermentation-derived short-chain fatty acids such as butyrate that interact with the nicotinic acid receptor [139, 140]. This flip also relates to the altered education of immune systems as "Absent Old Friends" affect the differentiation and migration of antigen-specific protective regulatory T cells and the balance with pro-inflammatory T helper 17 (with BCG having mitigating effects). The result is "immune intolerance" to otherwise harmless antigens and allergic and auto-immune disease [141, 142]. As already mentioned a prequel took place in the Spanish New World when those locals on a higher meat diet developed "sneezes".

27. So long so much auto-immunity – example of MS

Less Tryptophan in diet abrogates pathology in models of multiple sclerosis. MS is not the only auto-immune disease where one can link diet, microbiomes, autoreactive T cells, and IDO- 1 mediated tryptophan breakdown [143, 144]. Risk factors include meat, low Vitamin D, genetic pro-inflammatory predispositions, and inter-current infections that all affect T cell regulation. Adjusting tryptophan and nicotinamide in diet could lead to more resilient Treg/T (17) helper cell ratio – the same mechanism that stem-cells or the adoptive transfer of regulatory T cells, helminths or microbiomes are thought to work [145, 146].

28. Modern diseases and the ageing stakes – highs and lows

NNMT is a detoxification enzyme reducing nicotinamide levels that controls behaviour, neurodegeneration and lifespan by regulating energy, methylome and autophagy. NNMT is raised in many diseases of affluence whilst NAD levels fall: enzyme induction could be from high nicotinamide intake [147, 148] (**Figure 5**). As Brenner has said "NAD coenzymes catalyse the conversion of everything we eat in to everything we are and everything we do". High nicotinamide dosage from plentiful meat and milk often with supplements may play a part in diseases of affluence as is fairly well established for

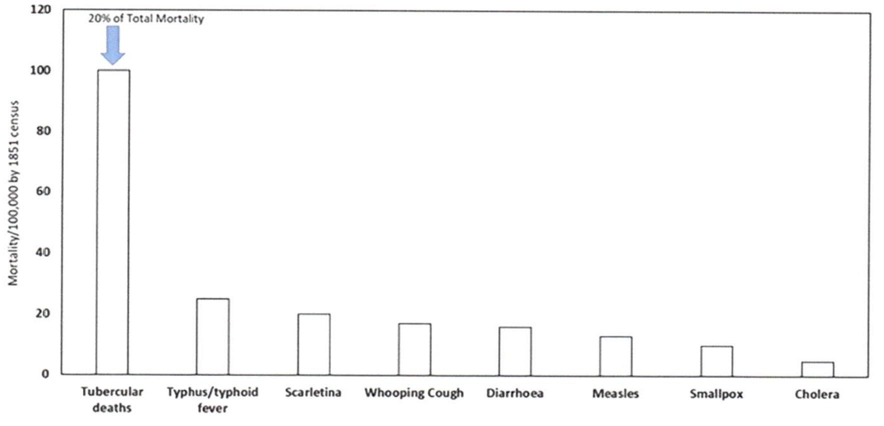

Figure 4.
TB, the "White Death," mortality shown using London data for 1850, TB vanished as meat intake increased - chiefly from imports (in exchange for cotton goods) that in effect exported infectious diseases to the poorly fed and low meat tropics.

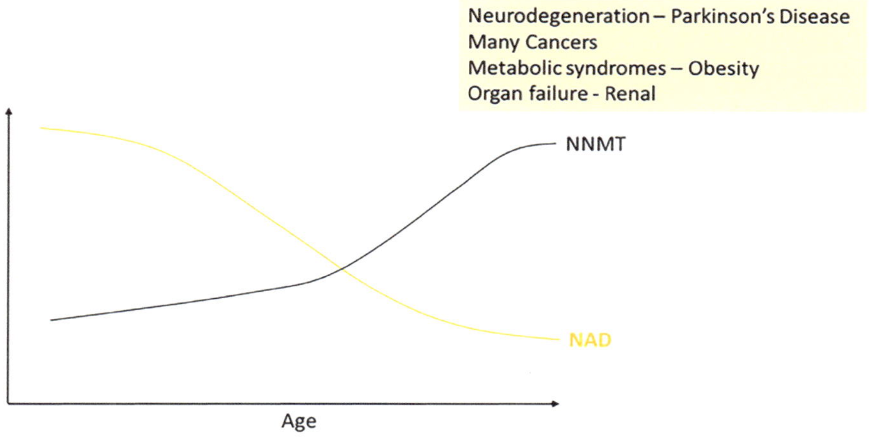

Figure 5.
NAD declines with age whereas NNMT levels rise in affluent geographies. Amongst the poor NAD levels would be low at all ages. Major preventive windows of opportunity present themselves for both rich and poor.

red or processed meat and cancer, particularly colorectal, and deaths and yet in Japan a "Goldilocks" diet with more meat and dairy is thought to be responsible for a decline in cerebrovascular mortality and their unusual longevity [149].

29. Pellagra: longevity at a price

Theories on ageing involve nicotinamide: pellagra was a real world case of premature ageing consistent with rises in life expectancy and lower incidence of dementia when diet improves [150, 151]. Longevity pathways, are activated by NAD booster molecules. NAD- rhythms are lynch-pins that explain circadian clocks and physiological states from hunger to fatigue to stress, and even the effects of alcohol. Antagonistic pleiotropy, a popular theory for ageing with genes important in development having adverse effects from relaxed selection in later life or developmental run-on includes NAD-consumer and NNMT genes [152, 153].

Pellagra comprised of dozens of mimics of neurodegenerative diseases and psychopathology that selectively affect high energy neurones in complex synaptic circuits. Topical explanations invoke proteinopathies, mitochondrial failure, inflammation, oxidant stress, calcium dysregulation, gut dysbioses, and neurotransmitter loss that were downstream events in pellagra [154, 155].

NAD may be the common denominator and "silver bullet" for cells with competing "mouths to feed" that with genetic or co-existent environmental factors gets channelled to various phenotypes spreading in "vulnerability networks" and prion-like waves. Nicotinamide may need to be adjusted by genome and age to avoid DOHaD, "disposable soma" or antagonistic pleiotropic effects that may only kick-in later in life requiring a higher nicotinamide and the more ancestral diet [156].

30. Nurture over nature: NAD World – barometers and monitors

Measuring ourselves embedded in an "NAD World" may be a parsimonious way of emancipating metabolic controls and energy flows to "refresh parts others cannot reach" by optimising nicotinamide dosage [157, 158] (**Figure 6**). Nicotinamide replacement or "Nutraceuticals" in general (often selling "candy" and empty calorie-ization) should not be the sole focus given negative effects on the methylome. Randomised trials varying meat intake are not realistic (first suggested by Daniel at the court of Nebuchadnezzar) but the predicted value, with a low ceiling effect, would lie in better cognition, resistance to microbes and "K" style fertility prioritising quality.

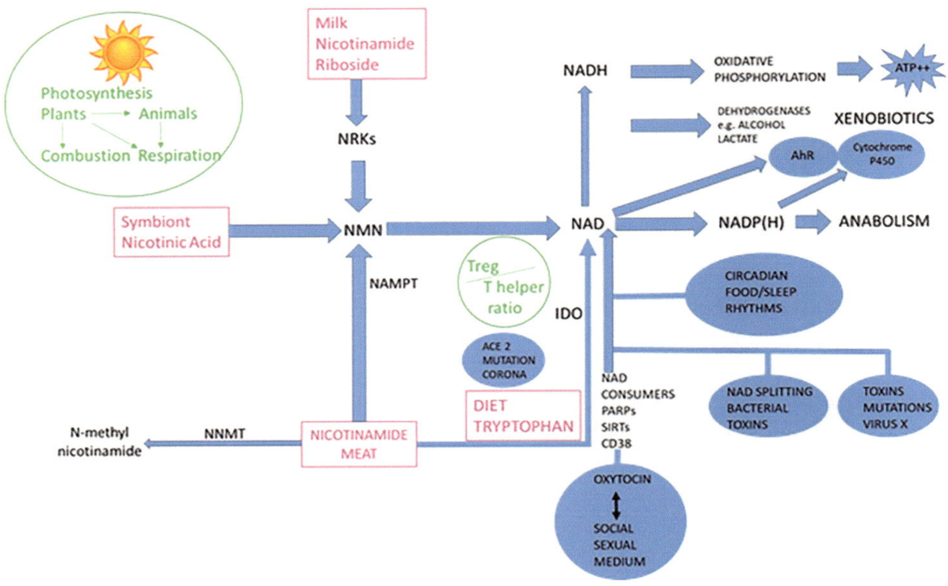

Figure 6.
This version of an "NAD World" has the dietary and social milieu, symbionts and pathogens all interacting with biochemical internal affairs. NAD has a "finger in every pie" affecting circadian rhythms, appetite and exercise alongside detoxification pathways for plant (and now drug) toxins with oxidant and other shocks from microbial pathogens and viruses that require resistance and (DNA) repair. Abbreviations: NMN=Nicotinamide mononucleotide; NAMPT = Nicotinamide phosphoribosyl-transferase; IDO = Indoleamine 2,3-dioxygenase; NNMT = Nicotinamide N-methyl-transferase; NRK = Nicotinamide riboside-kinase; PARP=Poly ADP-ribose polymerases; SIRTs = Sirtuins; CD38 = Cyclic ADP ribose-hydrolase; AhR = Aryl hydrocarbon receptor.

31. Human right to respire right

Subpar NAD levels are metabolic headwinds and pseudo-hypoxic states literally taking peoples "breath away" but, unlike meat, oxygen is free. Water is critical as splitting it is at the photosynthetic heart of an NAD World with riparian "hydraulic societies" raising civilizations [159, 160]. Although water can be a flashpoint on the whole cooperation has prevailed (with some high profile exceptions around dams or privatization), as it did over cleaning up water supplies to avoid infections such as cholera - perhaps because it was more obvious that the poor could infect the rich as is also true of air pollution (that now includes rising CO2) [161]. This danger is just as true for diet where obstacles should be overcome to deliver a "nicotinamide rush" as the platform for human capital, capacities and capabilities and to reduce the danger of zoonotic pandemics [162, 163].

32. Meat dangers: "X" diseases, "Y" plagues and zoonoses

Desperation for meat and cannibalism is implicated in prion diseases as feeding meat to herbivores triggered bovine spongiform encephalopathy and new version Jacob-Creutzfeld disease where NAD depletion has been implicated, consistent with the prion mimics seen in pellagra epidemics [164, 165].

33. Red flags and blind eyes: something new under the Sun

Opportunistic zoonoses are prominent (70%) causes of human scourges, a price of the (peri-) domestication of animals [166, 167]. Some think influenza strains and plagues arose and spread in tribes wandering with cattle over lands conquered by Genghis Khan [168, 169]. Recent emergent diseases include Marburg (1967), Ebola (1976), HIV (1981), Nipah (1998), SARS (2003) and other Coronaviruses like COVID-19 [170–172]. Cauldrons and hot-spots of emergent infections are built in high density populations with land cleared for agriculture encroaching on animal territories or are due to the desire for exotic foods [173, 174]. Those that heap opprobrium on current animal markets need to look back to London's 19th century costermongers who sold live meat in carnivalesque markets [175–178].

Poor and dangerous meat supplies have been described as "Structural violence" as for several billion wildlife consumption, or the income from household farming outside industrial "dragonhead" enterprises, is the only way of avoiding the "hidden hunger" of micronutrient deficiencies whether iron or vitamins A, D, B12 and B3 [179–182]. Campaigns to ban wildlife hunting needs thought if aimed to improve pandemic preparedness without leading to an even poorer diet for the "have-nots". As Lederberg said of viruses this is really a matter of *"Our Wits and their Genes"*. Zoonoses can be predicted and could be prevented by stringent surveillance of wildlife consumption with safe-guards including better hygiene with butchers and less exposure of Guano farmers to bat droppings [183, 184].

34. COVID-19 exposes an achilles heel

Pathogenic coronaviruses use the inducible angiotensin converting enzyme (ACE2) receptor to invade species that has roles in renin-aldosterone, tryptophan, immune-competence, and the microbiome [185, 186]. ACE2 is a chaperone for the

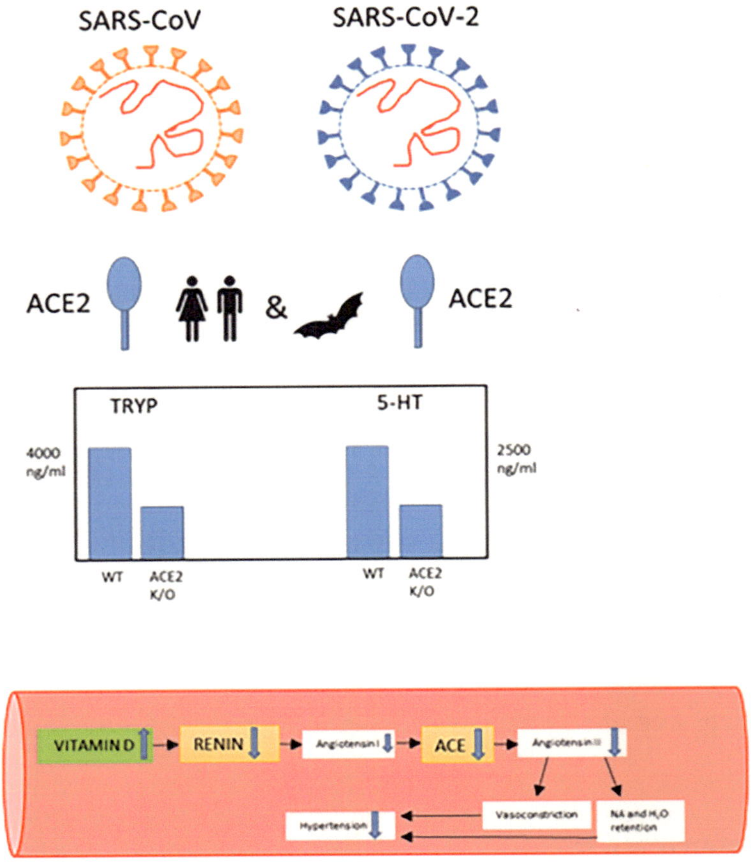

Figure 7.
Pleiotropic ACE-2 receptor and some overlooked interactions. ACE-2 affects Tryptophan uptake and the BoAT1 neutral amino-acid system and therefore the kynurenine and the T cell and interferon dependant "immune tolerance" pathway and exacerbates lost NAD homeostasis from pre-existing conditions (such as age or poverty and poor diet) or infection induced oxidative stress and its repair. Coronaviruses could, like ACE-2 knock-downs or the BoAT1 mutations that lead to the Hartnup pellagrous phenotype, reduce tryptophan and therefore serotonin levels and cause pellagra-like symptomatology both in the acute phase and as "long Covid" if not corrected. The renin-angiotensin system also involved in the pathophysiology could be affected by other vitamins such as Vitamin D.

amino acid transporter particularly regulating tryptophan uptake and interacting with Hartnup mutations that cause a multifactorial pellagra-like disorder. Metabolomics suggests that Covid-19 may have similar effects to *ace*2 knockouts affecting tryptophan convoys with loss of T cell homeostasis and Interferon responses affecting reactive and over-reactive immune responses [187, 188] (**Figure 7**). Some effective Covid treatments such as Dexamethasone and Tocilizumab affect this kynurenine pathway [189, 190]. Prominent enteritis and neuropsychiatric complications with (myoclonic) encephalopathy and "Long Covid" and other delayed complications are reminiscent of pellagra. As with other microbes being NAD-replete in the first place should improve host resistance and low initial NAD levels may explain several risk factors such as age, poverty and disability particularly if then exacerbated by post-Covid austerity diets as economies fail [141, 191–193].

35. Population matters redux – crunch-time for non-coercive measures

Earlier we referred to Malthus' observations on a cereal dependant population and introduced meat in to the demographic debate as a quality versus quantity piece of a complex jig-saw [194–196]. Formulae such as Environmental impact = Population x Energy consumed per capita - show that population counts particularly when energy consumed per person is high [197, 198]. Coercive population measures have had mixed results as have state "cash for babies" procreation policies and has stigmatised debate. Cereal supplements increase infant birth weight but reduce time to next pregnancy whereas a diet with adequate meat directly and indirectly (through better education) speeds demographic transitions.

The extremes are striking with population predicted to fall by 50% in rich countries but to increase by 300% in poor African nations, such as "zestful" Nigeria, with consequences for age structure, economic potential, migration and geopolitical power. There is currently little recognition of dietary drivers even though de Castro proposed that malnutrition was the cause not the effect of low quality population explosions 70 years ago [199].

36. Cutting to the chase: mean about meat means to a bad end

A remedy is to retro-shift to the 18th C idea of liberty that imposes state obligations to ensure "*bon marche*" not basic "bread and circuses". Adam Smith wrote, after observing European induced injustices, "greater wealth may inspire respect for the rights of one another" with a fairer "slice of the pie". Peak meat has surely passed for the rich and needs to be levelled at say 30 kg pa reducing food related emissions by a third (or more if switching from beef) and benefitting health. Given the world is home to 5 billion ungulates and 22 billion chickens this surely should provide an optimal "flexitarian" diet for all - with a role for plant-based meat substitutes and affordable lab-grown meat or tucking into "cricket snacks". Many political systems have accepted the need to supply grain yet none treat meat as a need rather than for those who have the means (16th C Henri IV of France's "chicken in the pot" for peasants was the exception). Rulers, from fascists to socialists, have recognised the power of food as a tool for their territorial ambitions whilst not balking at using it to starve their own people or only supporting equal sustenance for the working classes if linked to productivity [200]. One predicament of modern democracies is that they legitimize and spend large sums on defined disease, much in the last years of life, yet delegitimize those in dietary poverty normalizing their premature deaths. As Kropotkin (1892) said "*Well-being for all should not be a dream*".

If regimes addressed these dietary issues rulers may find that their citizens are healthier and less likely to reject basic democratic principles or descend in to monoculturalism, restrictive immigration or insurrections.

The ill effects of inequality, austerity and pauperism from "*Ancien Regimes*" to modern times on health, well-being and social mobility are well documented. Most narratives swing between clashes between oppressors and the proud oppressed and how self-serving oligarchies have self-perpetuated. The exact mechanism for harm, other than invoking stress or "social determinants" or "weathering", is however unclear. Stress reduction is, after all, convincingly invoked as the reason for pyramids of power and hierarchy [201]. Here we spell out how this originally happened in line with ecological and metabolic rift observations on the later effects of industrialization allowing food meccas and food ghettos and deserts [2, 202]. We propose that once the meat supply became constrained, we evolved on a dietary spectrum with a high meat to cereal ratio supporting a ruling intellectual elite and a low ratio a fertile proletarian essential (yet disposable) working class - and when there is a surplus of population, an unsupported underclass prone to rebellion [48]. Turchin however also points out that elite overproduction and intra-elite competition in gilded ages (such as the 1920's and now) marked by extremes of income, height and health inequality has often preceded ages of discord and societal collapses before a more progressive new-deal social construct and ecological revival. Increased equality should proceed on a "de-growth" and socio-ecological agenda recognizing that there is an abundance of good food to be shared if better managed rather than acting as if the calorie-ization and empty calorie-ization of the poor has solved the problem rather than becoming a tangible commercial determinant of health adversely affecting NAD homeostasis. This drive for meat security is more sustainable than continuing with a scenario with an artificial scarcity of meat and other "luxuries" encouraged by capitalist concentrations of power in the mega-merged agri-food "Big Food" profit driven sector that leads to reduced public wealth but private riches, biodiversity loss and excess emissions [203, 204].

Meat elites are now redundant developmental over-runs (not unlike some theories of cancer). Affirmative action needs to correct this dietary discord or actions aimed at the facades fronting inequality will fail. The opposite of inequality in this context is not a Utopian state or a meritocracy but equity of provisions with better metabolic homeostasis and no NAD headwinds for the poor. Hinman and Harris (1939) recognised that the meat eating races and classes have been instrumental to progress and that meatification is a marker and the ladder of class ascension and social mobility. Reframing Aristotle, this corresponds to a hierarchy of needs with a physiologically good diet being met free as a public good (it basically is already free for the rich) but the equally important self-actualization wants for a good life being left more to an individual's freedom and drive. Redistributing quality food has been modelled from social and economic perspectives in a new "Moral Economy" as "Sitopias" and "Diets for a Small Planet" that could now be grounded in the constitution and currency of an "NAD World" - and seen more as an investment as it closes innovation gaps, as seen in China, as well as reducing risks from pandemics or "superbug" antibiotic resistance [205–209]. Families may be the place to start as they already have "Burkian" style covenants between the dead, the living and those yet to be born. Enough family income to provide meat reflected in more shapely Engel Curves locked in to a top-down international governance structure could work as a "Gramscian" common-sense counter-hegenomic bloc and cry from those stuck in the basement [210]. Gramsci's words ring true "The old is dying and the new cannot be born; in this interregnum a great variety of morbid symptoms appear".

37. Conclusion

As Thoreau suggested with capitalism and its attendant inequalities in mind *"If Icarian thoughts returned to ground would we go to heaven the long way round"*. Our solution speaks for an algorithm that opens secure and safe meat larders derived from agro-ecological farming regimes that respect the best of the organic and food sovereign movements without forsaking scientific or commercial approaches shorn of soil degeneration from high tillage, excess fertilisers and pesticides and mono-crops [211, 212]. New meat technology should help but at the least cleaned up meat production from grass-and even feed-lot grain based farms to table will help and are unlikely to become "stranded" assets any time soon [213].

Quixotic quests for preventive causes for every known complication of poverty could be avoided by moving the dial to find a "sweet-spot" to avoid nicotinamide under-and over-load. Fair reform could happen without imposing widespread vegetarianism – a vaunted solution that would not benefit the needs of the nicotinamide have-not-half. Discrimination, we say, piggy-backs on meat extremes and could dissolve as it did for the pellagra-ridden "Butterfly caste", with meat justice leading at the least to a new emergent chapter in the history of inequality by abolishing "Precariats and Proletariats". Black Egyptian educators were after all the sparks of modern Europe not the blonde races or the later Anglosphere. Condorcet (1795) divided history into ten periods, the last of which permitted *"the abolition of inequality between nations, the progress of equality within each nation, and the true perfection of mankind".* A global overhaul that enables NAD equity would return us to our "other regarding" and sharing roots that, after some detours to boost population, began with meat and land equality - as well as showing that we can rise above Kant's "self-incurred immaturity" in a new enlightenment movement with values that this time round are fair to all and might solve a more general syndemic crisis.

Funding

The authors disclosed financial support from the QEHB Charity.

Declaration of conflicting interests

The author(s) declared no conflicts of interest with respect to the research, authorship, or publication of this article.

Author details

Adrian C. Williams[1]* and Lisa J. Hill[2]

1 Department of Neurology, University Hospitals Birmingham NHS Foundation Trust, Birmingham, UK

2 School of Biomedical Sciences, Institute of Clinical Sciences, University of Birmingham, Birmingham, UK

*Address all correspondence to: adrian.williams1949@gmail.com

References

[1] Prohaska A, Racimo F, Schork AJ, Sikora M, Stern AJ, Ilardo M, Allentoft ME, Folkersen L, Buil A, Moreno-Mayar JV. Human disease variation in the light of population genomics. Cell. 2019;177(1):115-131.

[2] Wells JCK. The Metabolic Ghetto: An Evolutionary Perspective on Nutrition, Power Relations and Chronic Disease: Cambridge University Press; 2016.

[3] Atkinson AB. Inequality: What can be done?: Harvard University Press; 2015.

[4] Kohler TA, Smith ME. Ten Thousand Years of Inequality: The Archaeology of Wealth Differences: University of Arizona Press; 2018.

[5] Milanovic B. Global Inequality: A New Approach for the Age of Globalization: Harvard University Press; 2016.

[6] O'Connor C. The Origins of Unfairness: Social Categories and Cultural Evolution: Oxford University Press; 2019.

[7] Norberg J. Open: The Story of Human Progress: Atlantic Books; 2020.

[8] Sachs JD. The Ages of Globalization: Geography, Technology, and Institutions: Columbia University Press; 2020.

[9] Williams AC, Hill LJ. The 4 D's of Pellagra and Progress. International Journal of Tryptophan Research. 2020;13:1178646920910159.

[10] Williams AC, Ramsden DB. Pellagra: A clue as to why energy failure causes diseases? Med Hypotheses. 2007;69(3):618-628.

[11] Prais SJ, Houthakker HS. The Analysis of Family Budgets: Cambridge University Press; 1971.

[12] Bello SM, Wallduck R, Dimitrijevic V, Zivaljevic I, Stringer CB. Cannibalism versus funerary defleshing and disarticulation after a period of decay: comparisons of bone modifications from four prehistoric sites. Am J Phys Anthropol. 2016;161(4):722-743.

[13] Cole J. Assessing the calorific significance of episodes of human cannibalism in the Palaeolithic. Sci Rep. 2017;7:44707.

[14] Barker F, Hulme P, BARKER FD, Iversen M, Hulme PLP, Arens WF, Obeyeskere G, Maduriera L, Huggan G, Kraniauskas J. Cannibalism and the Colonial World: Cambridge University Press; 1998.

[15] Lang T. Feeding Britain: Our Food Problems and How to Fix Them: Penguin Books Limited; 2020.

[16] McMahon P. Feeding Frenzy: The New Politics of Food: Profile; 2013.

[17] Harper S. Demography: A Very Short Introduction: Oxford University Press; 2018.

[18] Livi-Bacci M. Population and Nutrition: An Essay on European Demographic History: Cambridge University Press; 1991.

[19] Bashford A, Chaplin JE. The New Worlds of Thomas Robert Malthus: Rereading the Principle of Population: Princeton University Press; 2016.

[20] Eversley DEC. Social Theories of Fertility and the Malthusian Debate: Greenwood Press; 1975.

[21] Jasienska G, Bribiescas RG, Furberg AS, Helle S, Nunez-de la Mora A. Human reproduction and health: an evolutionary perspective. Lancet. 2017;390(10093):510-520.

[22] Kaptijn R, Thomese F, Liefbroer AC, Van Poppel F, Van Bodegom D, Westendorp RGJ. The Trade-Off between Female Fertility and Longevity during the Epidemiological Transition in the Netherlands. PLoS One. 2015;10(12):e0144353.

[23] Ear PH, Chadda A, Gumusoglu SB, Schmidt MS, Vogeler S, Malicoat J, Kadel J, Moore MM, Migaud ME, Stevens HE, et al. Maternal Nicotinamide Riboside Enhances Postpartum Weight Loss, Juvenile Offspring Development, and Neurogenesis of Adult Offspring. Cell Rep. 2019;26(4):969-983.e964.

[24] Nenko I, Hayward AD, Simons MJP, Lummaa V. Early-life environment and differences in costs of reproduction in a preindustrial human population. PLoS One. 2018;13(12):e0207236.

[25] Dyson T. Population and Development: The Demographic Transition: Zed Books; 2013.

[26] North DC, Thomas RP. The Rise of the Western World: A New Economic History: Cambridge University Press; 1973.

[27] Dartnell L. Origins: How The Earth Made Us: Random House; 2019.

[28] Lemke A. Foraging in the Past: Archaeological Studies of Hunter-Gatherer Diversity: University Press of Colorado; 2019.

[29] Deary IJ. Intelligence: A very short introduction: Oxford University Press; 2020.

[30] Kaplan H, Hill K, Lancaster J, Hurtado AM. A theory of human life history evolution: diet, intelligence, and longevity. Evolutionary Anthropology: Issues, News, and Reviews. 2000;9(4):156-185.

[31] Downey SS, Haas WR, Jr., Shennan SJ. European Neolithic societies showed early warning signals of population collapse. Proc Natl Acad Sci U S A. 2016;113(35): 9751-9756.

[32] Mellars P, French JC. Tenfold population increase in western europe at the neandertal–to–modern human transition. Science. 2011;333(6042): 623-627.

[33] Harris S. What Have Plants Ever Done for Us?: Western Civilization in Fifty Plants: Bodleian Library; 2015.

[34] Le S. 100 Million Years of Food: What Our Ancestors Ate and Why It Matters Today: Picador; 2016.

[35] El Zaatari S, Grine FE, Ungar PS, Hublin J-J. Neandertal versus modern human dietary responses to climatic fluctuations. PLoS One. 2016;11(4):e0153277.

[36] Page AE, Chaudhary N, Viguier S, Dyble M, Thompson J, Smith D, Salali GD, Mace R, Migliano AB. Hunter-gatherer social networks and reproductive success. Sci Rep. 2017;7(1):1153.

[37] Hood B. The Domesticated Brain: A Pelican Introduction: Penguin Books Limited; 2014.

[38] Maslow AH. Toward a Psychology of Being: Start Publishing LLC; 2013.

[39] Suzman J. Work: A History of How We Spend Our Time: Bloomsbury Publishing; 2020.

[40] Eaton SB, Konner M. Paleolithic nutrition. A consideration of its nature and current implications. N Engl J Med. 1985;312(5):283-289.

[41] Rutledge GA, Cabral LG, Kuey BJ, Lee JD, Mueller LD, Rose MR. Hamiltonian patterns of age-dependent adaptation to novel environments. PLoS One. 2020;15(10):e0240132.

[42] Fuller DQ, Stevens CJ. Between domestication and civilization: the role of agriculture and arboriculture in the emergence of the first urban societies. Veg Hist Archaeobot. 2019;28(3): 263-282.

[43] Scott JC. Against the Grain: A Deep History of the Earliest States: Yale University Press; 2017.

[44] Blake M. Maize for the Gods: Unearthing the 9,000-Year History of Corn: University of California Press; 2015.

[45] Miller G. The Mating Mind: How Sexual Choice Shaped the Evolution of Human Nature: Vintage; 2001.

[46] Collingham L. The Biscuit: The History of a Very British Indulgence: Random House; 2020.

[47] DiAngelo R. White Fragility: Why It's So Hard for White People to Talk About Racism: Penguin Books Limited; 2019.

[48] Kara S. Modern Slavery: A Global Perspective: Columbia University Press; 2017.

[49] Scheidel W. The Great Leveler: Violence and the History of Inequality from the Stone Age to the Twenty-First Century: Princeton University Press; 2018.

[50] Goody J, Goody JR, Press CU, Dunn J, Hawthorn G. Cooking, Cuisine and Class: A Study in Comparative Sociology: Cambridge University Press; 1982.

[51] Jones M. Feast: Why Humans Share Food: OUP Oxford; 2008.

[52] Amorim CEG, Vai S, Posth C, Modi A, Koncz I, Hakenbeck S, La Rocca MC, Mende B, Bobo D, Pohl W, et al. Understanding 6th-century barbarian social organization and migration through paleogenomics. Nat Commun. 2018;9(1):3547.

[53] Piketty T. Capital in the 21st Century. 2014.

[54] Nielsen R, Akey JM, Jakobsson M, Pritchard JK, Tishkoff S, Willerslev E. Tracing the peopling of the world through genomics. Nature. 2017;541(7637):302-310.

[55] Reich D. Who We Are and How We Got Here: Ancient DNA and the New Science of the Human Past: Oxford University Press; 2018.

[56] Harper K. The Fate of Rome: Climate, Disease, and the End of an Empire: Princeton University Press; 2017.

[57] Hansen V. The Year 1000: When Explorers Connected the World – and Globalization Began: Penguin Books Limited; 2020.

[58] Moore RI. The First European Revolution: 970-1215: Wiley; 2000.

[59] Cohn SK. The Black Death Transformed: Disease and Culture in Early Renaissance Europe: Arnold; 2002.

[60] Kohler TA, Smith ME, Bogaard A, Feinman GM, Peterson CE, Betzenhauser A, Pailes M, Stone EC, Marie Prentiss A, Dennehy TJ, et al. Greater post-Neolithic wealth disparities in Eurasia than in North America and Mesoamerica. Nature. 2017;551(7682): 619-622.

[61] Sombart W, Atterbury AP. Socialism and the Social Movement in the 19th Century: Creative Media Partners, LLC; 2018.

[62] Crosby AW. The Columbian exchange: biological and cultural consequences of 1492: Greenwood Publishing Group; 2003.

[63] Richards JF. The Unending Frontier: An Environmental History of the Early Modern World: University of California Press; 2005.

[64] Colas A. Food, Politics, and Society: Social Theory and the Modern Food System: University of California Press; 2018.

[65] Collingham EM. Imperial Bodies: The Physical Experience of the Raj, C.1800-1947: Wiley; 2001.

[66] Earle R. The Body of the Conquistador: Cambridge University Press; 2014.

[67] MAZUMDAR S. The Impact of New World Food Crops on the Diet and Economy of China and India, 1600-1900. IN GREW, R.(Ed.) Food in Global History. Boulder, Colorado. Westview Press; 1999.

[68] Warman A. Corn and capitalism: How a botanical bastard grew to global dominance: Univ of North Carolina Press; 2003.

[69] Grace D, Lindahl J, Wanyoike F, Bett B, Randolph T, Rich KM. Poor livestock keepers: ecosystem-poverty-health interactions. Philos Trans R Soc Lond B Biol Sci. 2017;372(1725).

[70] Collared M, Kemery M, Banks S. Causes of toolkit variation among hunter-gatherers: a test of four competing hypotheses. Canadian Journal of Archaeology/Journal Canadien D'Archéologie. 2005:1-19.

[71] Langer WL. Europe's initial population explosion. The American Historical Review. 1963;69(1):1-17.

[72] McNeill W. Plagues and Peoples: Knopf Doubleday Publishing Group; 2010.

[73] Bohstedt J. The Politics of Provisions: Food Riots, Moral Economy, and Market Transition in England, c. 1550-1850: Taylor & Francis; 2016.

[74] Midlarsky MI. The Evolution of Inequality: War, State Survival, and Democracy in Comparative Perspective: Stanford University Press; 1999.

[75] Jablonski NG, Chaplin G. The colours of humanity: the evolution of pigmentation in the human lineage. Philos Trans R Soc Lond B Biol Sci. 2017;372(1724).

[76] Quillen EE, Norton HL, Parra EJ, Lona-Durazo F, Ang KC, Illiescu FM, Pearson LN, Shriver MD, Lasisi T, Gokcumen O, et al. Shades of complexity: New perspectives on the evolution and genetic architecture of human skin. Am J Phys Anthropol. 2019;168 Suppl 67:4-26.

[77] Du Bois Center at the University of Massachusetts TWEB, Battle-Baptiste W, Rusert B. W. E. B. Du Bois's Data Portraits: Visualizing Black America: Princeton Architectural Press; 2018.

[78] Green T. A Fistful of Shells: West Africa from the Rise of the Slave Trade to the Age of Revolution: Penguin Books Limited; 2019.

[79] Fogel RW, Engerman SL. Time on the Cross: The Economics of American Negro Slavery: Norton; 1995.

[80] Daniel P. Breaking the Land: The Transformation of Cotton, Tobacco, and Rice Cultures Since 1880: University of Illinois Press; 1986.

[81] Graham LO. Our Kind of People: Inside America's Black Upper Class: HarperCollins e-books; 2009.

[82] Estes N. Our History Is the Future: Standing Rock Versus the Dakota Access Pipeline, and the Long Tradition of Indigenous Resistance: Verso Books; 2019.

[83] DINER HR. Hungering for America: Harvard University Press; 2009.

[84] Harrington JR, Gelfand MJ. Tightness–looseness across the 50 united states. Proceedings of the National Academy of Sciences. 2014;111(22):7990-7995.

[85] Talhelm T, English AS. Historically rice-farming societies have tighter social norms in China and worldwide. Proceedings of the National Academy of Sciences. 2020;117(33):19816-19824.

[86] Shilliam R. Race and the Undeserving Poor: From Abolition to Brexit: Agenda Publishing; 2018.

[87] Levine S. School Lunch Politics: The Surprising History of America's Favorite Welfare Program: Princeton University Press; 2011.

[88] Appelbaum R. Aguecheek's Beef, Belch's Hiccup, and Other Gastronomic Interjections: Literature, Culture, and Food Among the Early Moderns: University of Chicago Press; 2008.

[89] Spang RL. The Invention of the Restaurant: Paris and Modern Gastronomic Culture: Harvard University Press; 2020.

[90] Ash TG. Polish Revolution: HarperCollins Publishers Limited; 1998.

[91] Churchwell S. Behold, America: A History of America First and the American Dream: Bloomsbury Publishing; 2018.

[92] Johnson W. River of Dark Dreams: Harvard University Press; 2013.

[93] Manne K. Entitled: How Male Privilege Hurts Women: Penguin Books Limited; 2020.

[94] Sperling G. Economic Dignity: Penguin Publishing Group; 2020.

[95] Wilkerson I. Caste: The International Bestseller: Penguin Books Limited; 2020.

[96] Collins PH, Bilge S. Intersectionality: Wiley; 2020.

[97] Henrich J. The Weirdest People in the World: How the West Became Psychologically Peculiar and Particularly Prosperous: Penguin Books Limited; 2020.

[98] Sowell T. Intellectuals and Society: Basic Books; 2012.

[99] Ellis EC, Ellis EC. Anthropocene: A Very Short Introduction: Oxford University Press; 2018.

[100] Piperno DR, McMichael C, Bush MB. Amazonia and the Anthropocene: What was the spatial extent and intensity of human landscape modification in the Amazon Basin at the end of prehistory? The Holocene. 2015;25(10):1588-1597.

[101] Fagan B. The Little Ice Age: How Climate Made History 1300-1850: Basic Books; 2019.

[102] Lewis SL, Maslin MA. The Human Planet: How We Created the Anthropocene: Penguin Books Limited; 2018.

[103] Steffen W, Rockstrom J, Richardson K, Lenton TM, Folke C, Liverman D, Summerhayes CP, Barnosky AD, Cornell SE, Crucifix M, et al. Trajectories of the Earth System in the Anthropocene. Proc Natl Acad Sci U S A. 2018;115(33):8252-8259.

[104] Caparros M. Hunger: The Oldest Problem: Melville House; 2020.

[105] Howlett J. An examination of Dr. Price's Essay on The Population of England and Wales London. 1781;91.

[106] Linebaugh P. Stop, Thief!: The Commons, Enclosures, and Resistance: PM Press; 2014.

[107] Macfarlane A. The Origins of English Individualism: The Family Property and Social Transition: Wiley; 1978.

[108] Davis M. Late Victorian Holocausts: El Nino Famines and the Making of the Third World: Verso Books; 2002.

[109] Weaver JC. Great Land Rush and the Making of the Modern World, 1650-1900: MQUP; 2003.

[110] Stoler AL. Imperial Debris: On Ruins and Ruination: Duke University Press; 2013.

[111] Wylie D. Starving on a Full Stomach: Hunger and the Triumph of Cultural Racism in Modern South Africa: University Press of Virginia; 2001.

[112] Taylor M. The Interest: How the British Establishment Resisted the Abolition of Slavery: Random House; 2020.

[113] Timmins N. The Five Giants [New Edition]: A Biography of the Welfare State: HarperCollins Publishers; 2017.

[114] Pilcher JM. Food in World History: Taylor & Francis; 2017.

[115] Lindert PH, Williamson JG. Unequal Gains: American Growth and Inequality since 1700: Princeton University Press; 2017.

[116] Putnam RD, Garrett SR. The Upswing: How America Came Together a Century Ago and How We Can Do It Again: Simon & Schuster; 2020.

[117] Hennessy MB, Deak T, Schiml PA. Sociality and sickness: have cytokines evolved to serve social functions beyond times of pathogen exposure? Brain Behav Immun. 2014;37:15-20.

[118] Watts SJ. Epidemics and History: Disease, Power, and Imperialism: Yale University Press; 1999.

[119] Coogan TP. The Famine Plot: England's Role in Ireland's Greatest Tragedy: St. Martin's Publishing Group; 2012.

[120] Johnston BF, Mellor JW. The role of agriculture in economic development. The American Economic Review. 1961;51(4):566-593.

[121] Zotor FB, Ellahi B, Amuna P. Applying the food multimix concept for sustainable and nutritious diets. Proc Nutr Soc. 2015;74(4):505-516.

[122] Easterly W, Easterly WR. The White Man's Burden: Why the West's Efforts to Aid the Rest Have Done So Much Ill and So Little Good: Penguin Press; 2006.

[123] Getachew A. Worldmaking After Empire: The Rise and Fall of Self-Determination: Princeton University Press; 2020.

[124] Arboleda M. Planetary Mine: Territories of Extraction under Late Capitalism: Verso Books; 2020.

[125] Rodney W, Davis A. How Europe Underdeveloped Africa: Verso Books; 2018.

[126] Caballero B. The Nutrition Transition: Diet and Disease in the Developing World: Elsevier Science; 2002.

[127] Marcus E. Meat Market: Animals, Ethics, & Money: Brio Press; 2005.

[128] Perren R. The Meat Trade in Britain, 1840-1914. London: Routledge and Kegan Paul; 1978.

[129] de Castro J. Geography of Hunger: Gollancz; 1952.

[130] Peffer N. The White Man's Dilemma: Climax of the Age of Imperialism: John Day Company; 1927.

[131] Harrison M. Disease and the Modern World: 1500 to the Present Day: Wiley; 2013.

[132] Porter R. The Greatest Benefit to Mankind: A Medical History of Humanity (The Norton History of Science): W. W. Norton; 1999.

[133] Mc KD, Malone L, et al. The effect of nicotinic acid amide on experimental tuberculosis of white mice. J Lab Clin Med. 1948;33(10):1249-1253.

[134] Simmons JD, Peterson GJ, Campo M, Lohmiller J, Skerrett SJ, Tunaru S, Offermanns S, Sherman DR, Hawn TR. Nicotinamide limits replication of *Mycobacterium tuberculosis* and BCG within macrophages. J Infect Dis. 2019.

[135] Henkel JS, Baldwin MR, Barbieri JT. Toxins from bacteria. Molecular, Clinical and Environmental Toxicology: Springer; 2010. p. 1-29.

[136] Mahon RN, Hafner R. Immune Cell Regulatory Pathways Unexplored as Host-Directed Therapeutic Targets for *Mycobacterium tuberculosis*: An Opportunity to Apply Precision Medicine Innovations to Infectious Diseases. Clin Infect Dis. 2015;61Suppl 3:S200-216.

[137] Daschner A, Gonzalez Fernandez J. Allergy in an Evolutionary Framework. J Mol Evol. 2020;88(1):66-76.

[138] Trowell HC, Burkitt DP. Western Diseases, Their Emergence and Prevention. London: Edward Arnold; 1981.

[139] Munn DH, Zhou M, Attwood JT, Bondarev I, Conway SJ, Marshall B, Brown C, Mellor AL. Prevention of allogeneic fetal rejection by tryptophan catabolism. Science. 1998;281(5380):1191-1193.

[140] Sonnenburg JL, Backhed F. Diet-microbiota interactions as moderators of human metabolism. Nature. 2016;535(7610):56-64.

[141] Hashimoto T, Perlot T, Rehman A, Trichereau J, Ishiguro H, Paolino M, Sigl V, Hanada T, Hanada R, Lipinski S, et al. ACE2 links amino acid malnutrition to microbial ecology and intestinal inflammation. Nature. 2012;487(7408):477-481.

[142] Spinelli P, Latchney SE, Reed JM, Fields A, Baier BS, Lu X, McCall MN, Murphy SP, Mak W, Susiarjo M. Identification of the novel Ido1 imprinted locus and its potential epigenetic role in pregnancy loss. Hum Mol Genet. 2019;28(4):662-674.

[143] Swank RL, Lerstad O, Strom A, Backer J. Multiple sclerosis in rural Norway its geographic and occupational incidence in relation to nutrition. N Engl J Med. 1952;246 (19):722-728.

[144] Taylor BV. The major cause of multiple sclerosis is environmental: genetics has a minor role--yes. Mult Scler. 2011;17(10):1171-1173.

[145] Correale J, Ysrraelit MC, Gaitan MI. Immunomodulatory effects of Vitamin D in multiple sclerosis. Brain. 2009;132(Pt 5):1146-1160.

[146] Wekerle H. Brain Autoimmunity and Intestinal Microbiota: 100 Trillion Game Changers. Trends Immunol. 2017;38(7):483-497.

[147] Clement J, Wong M, Poljak A, Sachdev P, Braidy N. The Plasma NAD(+) Metabolome Is Dysregulated in "Normal" Aging. Rejuvenation Res. 2018.

[148] Neelakantan H, Brightwell CR, Graber TG, Maroto R, Wang HL, McHardy SF, Papaconstantinou J, Fry CS, Watowich SJ. Small molecule nicotinamide N-methyltransferase inhibitor activates senescent muscle stem cells and improves regenerative capacity of aged skeletal muscle. Biochem Pharmacol. 2019.

[149] Wolk A. Potential health hazards of eating red meat. J Intern Med. 2017;281(2):106-122.

[150] Jasienska G. Reproduction and lifespan: Trade-offs, overall energy budgets, intergenerational costs, and costs neglected by research. American Journal of Human Biology. 2009;21(4):524-532.

[151] Westendorp RG. Are we becoming less disposable?: Evolution has programmed us for early survival and reproduction but has left us vulnerable to disease in old age. In our present affluent environment, we are better adapting to these improved conditions. EMBO reports. 2004;5(1):2-6.

[152] Austad SN, Hoffman JM. Is antagonistic pleiotropy ubiquitous in aging biology? Evol Med Public Health. 2018;2018(1):287-294.

[153] Kanakkanthara A, Kurmi K, Ekstrom TL, Hou X, Purfeerst ER, Heinzen EP, Correia C, Huntoon CJ, O'Brien D, Wahner Hendrickson AE, et al. BRCA1 Deficiency Upregulates NNMT, Which Reprograms Metabolism and Sensitizes Ovarian Cancer Cells to Mitochondrial Metabolic Targeting Agents. Cancer Res. 2019;79(23): 5920-5929.

[154] Fu H, Hardy J, Duff KE. Selective vulnerability in neurodegenerative diseases. Nature neuroscience. 2018;21(10):1350-1358.

[155] Mattsson N, Schott JM, Hardy J, Turner MR, Zetterberg H. Selective vulnerability in neurodegeneration: insights from clinical variants of Alzheimer's disease. J Neurol Neurosurg Psychiatry. 2016;87(9):1000-1004.

[156] Hou Y, Lautrup S, Cordonnier S, Wang Y, Croteau DL, Zavala E, Zhang Y, Moritoh K, O'Connell JF, Baptiste BA. NAD+ supplementation normalizes key Alzheimer's features and DNA damage responses in a new AD mouse model with introduced DNA repair deficiency. Proceedings of the National Academy of Sciences. 2018:201718819.

[157] Elhassan YS, Philp AA, Lavery GG. Targeting NAD+ in metabolic disease; new insights into an old molecule. Journal of the Endocrine Society. 2017.

[158] Poddar SK, Sifat AE, Haque S, Nahid NA, Chowdhury S, Mehedi I. Nicotinamide Mononucleotide: Exploration of Diverse Therapeutic Applications of a Potential Molecule. Biomolecules. 2019;9(1).

[159] Cook GC. Disease and Sanitation in Victorian Britain: Lessons for the Third World: Melrose Books; 2015.

[160] Smith LC. Rivers of Power: How a Natural Force Raised Kingdoms, Destroyed Civilizations, and Shapes Our World: Penguin Books Limited; 2020.

[161] Stern M. Evidence that a mitochondrial death spiral underlies antagonistic pleiotropy. Aging Cell. 2017;16(3):435-443.

[162] Cochran G, Harpending H. The 10,000 Year Explosion: How Civilization Accelerated Human Evolution: Basic Books; 2009.

[163] Esping-Andersen G. The Three Worlds of Welfare Capitalism: Wiley; 2013.

[164] Haik S, Brandel JP. Infectious prion diseases in humans: cannibalism, iatrogenicity and zoonoses. Infect Genet Evol. 2014;26:303-312.

[165] Zhou M, Ottenberg G, Sferrazza GF, Hubbs C, Fallahi M, Rumbaugh G, Brantley AF, Lasmezas CI. Neuronal death induced by misfolded prion protein is due to NAD+ depletion and can be relieved in vitro and in vivo by NAD+ replenishment. Brain. 2015;138 (Pt 4):992-1008.

[166] Izcue A, Powrie F. Immunology: Malnutrition promotes rogue bacteria. Nature. 2012;487(7408):437-439.

[167] Morens DM, Folkers GK, Fauci AS. The challenge of emerging and re-emerging infectious diseases. Nature. 2004;430(6996):242-249.

[168] Dasgupta P, Ray D. Inequality as a Determinant of Malnutrition and Unemployment: Theory. The Economic Journal. 1986;96(384):1011-1034.

[169] Pearce-Duvet JM. The origin of human pathogens: evaluating the role of agriculture and domestic animals in the evolution of human disease. Biological Reviews. 2006;81(3):369-382.

[170] Andiman WA. Animals Viruses and Humans, A Narrow Divide: How Lethal Zoonotic Viruses Spill Over and Threaten Us: Paul Dry Books; 2018.

[171] Honigsbaum M. The Pandemic Century: A History of Global Contagion from the Spanish Flu to Covid-19: Ebury Publishing; 2020.

[172] Spinney L. Pale Rider: The Spanish Flu of 1918 and How it Changed the World: Random House; 2017.

[173] Valitutto MT, Aung O, Tun KYN, Vodzak ME, Zimmerman D, Yu JH, Win YT, Maw MT, Thein WZ, Win HH. Detection of novel coronaviruses in bats in Myanmar. PLoS One. 2020;15(4):e0230802.

[174] Volpato G, Fontefrancesco MF, Gruppuso P, Zocchi DM, Pieroni A. Baby pangolins on my plate: possible lessons to learn from the COVID-19 pandemic. Springer; 2020.

[175] Münch O. Henry Mayhew and the street traders of Victorian London—a cultural exchange with material consequences. The London Journal. 2018;43(1):53-71.

[176] Smith C. The wholesale and retail markets of London, 1660-1840. The Economic History Review. 2002;55(1): 31-50.

[177] Mayhew H. London Labour and the London Poor: A Cyclopaedia of the Condition and Earnings of Those that Will Work, Those that Cannot Work, and Those that Will Not Work: G. Woodfall; 1851.

[178] Winter J. London's Teeming Streets, 1830-1914: Taylor & Francis; 2013.

[179] Farmer P, Sen A, Sen M. Pathologies of Power: Health, Human Rights, and the New War on the Poor: University of California Press; 2005.

[180] Greenfeld KT. China Syndrome: The True Story of the 21st Century's First Great Epidemic: HarperCollins; 2007.

[181] Friant S, Ayambem WA, Alobi AO, Ifebueme NM, Otukpa OM, Ogar DA, Alawa CBI, Goldberg TL, Jacka JK, Rothman JM. Eating Bushmeat Improves Food Security in a Biodiversity and Infectious Disease "Hotspot". Ecohealth. 2020;17(1):125-138.

[182] Golden CD, Fernald LC, Brashares JS, Rasolofoniaina BR, Kremen C. Benefits of wildlife consumption to child nutrition in a biodiversity hotspot. Proceedings of the National Academy of Sciences. 2011;108(49):19653-19656.

[183] Kucharski A. The Rules of Contagion: Why Things Spread - and Why They Stop: Profile; 2020.

[184] Senthilingam M. Outbreaks and Epidemics: Battling infection from measles to coronavirus: Icon Books Limited; 2020.

[185] Hoffmann M, Kleine-Weber H, Schroeder S, Krüger N, Herrler T, Erichsen S, Schiergens TS, Herrler G, Wu N-H, Nitsche A. SARS-CoV-2 cell entry depends on ACE2 and TMPRSS2 and is blocked by a clinically proven protease inhibitor. Cell. 2020.

[186] Yan T, Xiao R, Lin G. Angiotensin-converting enzyme 2 in severe acute respiratory syndrome coronavirus and SARS-CoV-2: A double-edged sword? The FASEB Journal. 2020;34(5): 6017-6026.

[187] Alenina N, Bader M. ACE2 in brain physiology and pathophysiology: Evidence from transgenic animal models. Neurochem Res. 2019;44(6):1323-1329.

[188] Singer D, Camargo SM, Ramadan T, Schafer M, Mariotta L, Herzog B, Huggel K, Wolfer D, Werner S, Penninger JM, et al. Defective intestinal amino acid absorption in Ace2 null mice. Am J Physiol Gastrointest Liver Physiol. 2012;303(6):G686-695.

[189] Belladonna ML, Orabona C. Potential Benefits of Tryptophan Metabolism to the Efficacy of Tocilizumab in COVID-19. Frontiers in Pharmacology. 2020;11(959).

[190] Tian J, Zhang B, Rui K, Wang S. The Role of GITR/GITRL Interaction in Autoimmune Diseases. Front Immunol. 2020;11:588682-588682.

[191] Fischer DD, Kandasamy S, Paim FC, Langel SN, Alhamo MA, Shao L, Chepngeno J, Miyazaki A, Huang HC, Kumar A, et al. Protein Malnutrition Alters Tryptophan and Angiotensin-Converting Enzyme 2 Homeostasis and Adaptive Immune Responses in Human Rotavirus-Infected Gnotobiotic Pigs with Human Infant Fecal Microbiota Transplant. Clin Vaccine Immunol. 2017;24(8).

[192] Wallace R. Dead Epidemiologists: On the Origins of COVID-19: Monthly Review Press; 2020.

[193] Williams A. Adv Clin Neurosci Rehabil 2020; Pellagra: 4 D's and 8 Points. 2020.

[194] Williams AC, Hill LJ. Nicotinamide and Demographic and Disease transitions: Moderation is Best. International Journal of Tryptophan Research. 2019;12:1178646919855940.

[195] Kaufmann EP. Shall the Religious Inherit the Earth?: Demography and Politics in the Twenty-first Century: Profile Books; 2010.

[196] Pearson CS. On the Cusp: From Population Boom to Bust: Oxford University Press; 2015.

[197] Hartmann B. Reproductive Rights and Wrongs: The Global Politics of Population Control: Haymarket Books; 2016.

[198] Smil V. How many people can the earth feed? Population and Development Review. 1994:255-292.

[199] Collaborators GPaF. Population and fertility by age and sex for 195 countries and territories, 1950-2017: a systematic analysis for the Global Burden of Disease Study 2017. Lancet. 2018;392(10159):1995-2051.

[200] Rees L. Hitler and Stalin: The Tyrants and the Second World War: Penguin Books Limited; 2020.

[201] Perret C, Hart E, Powers ST. From disorganized equality to efficient hierarchy: how group size drives the evolution of hierarchy in human societies. Proc Biol Sci. 2020;287(1928): 20200693.

[202] Hanieh S, High H, Boulton J. Nutrition Justice: Uncovering Invisible Pathways to Malnutrition. Front Endocrinol (Lausanne). 2020;11:150.

[203] Finley E. Beyond the Limits of Nature: A Social-ecological Perspective on Degrowth as a Political Ideology. Capitalism Nature Socialism. 2019;30(2):244-250.

[204] Hickel J. Less is More: How Degrowth Will Save the World: Random House; 2020.

[205] Lowrey A. Give People Money: How a Universal Basic Income Would End Poverty, Revolutionize Work, and Remake the World: Crown; 2018.

[206] Patel R. The Value of Nothing: How to Reshape Market Society and Redefine Democracy: Portobello; 2011.

[207] Vollrath D. Fully Grown: Why a Stagnant Economy Is a Sign of Success: University of Chicago Press; 2020.

[208] Holt-Giménez E. Capitalism, food, and social movements: The political economy of food system transformation. Journal of Agriculture, Food Systems, and Community Development. 2019;9(A):23-35.

[209] Lappe FM, Lappé FM. Diet for a Small Planet: Ballantine Books; 1991.

[210] Vogler P. Scoff: A History of Food and Class in Britain: Atlantic Books; 2020.

[211] Denison RF. Darwinian Agriculture: How Understanding Evolution Can Improve Agriculture: Princeton University Press; 2016.

[212] Chai A, Moneta A. Retrospectives: engel curves. Journal of Economic Perspectives. 2010;24(1):225-240.

[213] Mellon J. Moo's Law: An Investor's Guide to the New Agrarian Revolution: Fruitful Publications; 2020.

Chapter 2

Cost Effectiveness of Poultry Production by Sustainable and Renewable Energy Source

Yuanlong Cui, Xuan Xue and Saffa Riffat

Abstract

Poultry farming is one of high energy consumption and energy-intensive industries that requires significant amount of fuel fossil to provide the desired internal temperature for health and production level of chicken, which results in high running cost and growth of greenhouse gas (GHG) emissions. Renewable and sustainable energy technologies are being employed in the area of poultry farming in order to achieve energy saving, GHG emission reduction and to some extent supply potential selective benefits for farmers. Therefore, it is very necessary for generalizing the state-of-the-art technologies including the solar photovoltaic, solar photovoltaic/thermal, ventilation and wind turbine, air/water/ground sources heat pump and thermal energy storage. It is demonstrated that the system energy saving could achieve up to 85% with a payback time of 3–8 years, compared to the conventional heating system.

Keywords: Poultry farm, Renewable and sustainable energy technology, Energy efficiency, Cost-effective, Payback period

1. Introduction

The ongoing coronavirus disease 2019 (COVID-19) global pandemic brought a few extreme challenges to the world including public health crisis, political, environmental, social and economic domains [1, 2]. Meanwhile, it reveals how population growth, globalization, urbanization, and mass travel give rise to a complex externality with far-reaching global impacts [3–5]. Notably, the COVID-19 has expounded the importance of addressing another global issue: global warming. The increasing atmospheric concentration of GHG is thought as the biggest contributor of global warming. This has exerted negative effect on plants, animals, human activity, ecosystems and economy around the world, which are largely associated to alterations in climate extremes [6]. Therefore, energy efficient and energy consumption saving have become more than important nowadays because of energy resource shortage, soaring energy prices as well as pressing environmental problems [7].

Notably, livestock production is responsible for GHG emission attaining 20–25% of global entire emission, of which approximately 70–80% stems from animal farm industries [8–10]. Poultry farm is an important sector that consumes large quantities of fuel all over the world. This is due to that the internal temperature, relative humidity, chemical environment, ventilation and lighting inside a chicken house would dramatically affect the growth of broilers, which should be kept within a reasonable scope [11]. The growth of broilers mainly depended on the internal environmental condition variation which may impede the meat and eggs production, such as heat or cold stress [12]. The desirable temperature and relative humidity requires to be kept between 26°C and 35°C and between 60% and 70%, respectively [13, 14]. The ammonia concentrations must be controlled below 25 ppm. This is because that high ammonia level may cause respiratory damages to the chicken [15]. Hence, the heating, cooling, lighting, temperature and ventilation need to be supervised accurately for better production [16, 17]. Traditional poultry farming seriously consume fossil fuel and gas via the power and heating systems for heating, cooling, lighting, ventilation and running electric motors for feed lines [18, 19]. Therefore, the usage of the renewable and sustainable energy technologies, including wind energy, solar energy, geothermal energy and air/water sources [20, 21], plays vital role on the poultry farm owing to their potential to a reduction of energy demand and welfare losses, economy and profitability, GHG reduction and conservation of resources [22–24]. Hence, in this chapter, these advanced technologies is investigated and summarized for easier tracking and better understanding of energy-efficient renovation for typical poultry houses.

2. Renewable and sustainable energy technologies for poultry farm

2.1 Solar energy technology

Solar energy is a very enormous, environment friendly and inexhaustible renewable energy resource. It is divided into solar photovoltaic (PV) technology, which convert the solar radiation into power generation, and solar thermal technology, which utilize the solar radiation directly for space heating, water heating, drying and cooking [25, 26].

2.1.1 Solar photovoltaic

Solar PV module is regarded as the electrical production element, and its performance is associated with the category and temperature of PV cell [27, 28]. When the PV cell temperature rises 1°C, the electrical conversion efficiency is reduced by approximately 0.4–0.5% for the crystalline silicon cell and about 0.25% for the amorphous silicon cell [29, 30]. Additionally, solar radiation is converted into direct current electricity by the PV module, thus is transformed into alternating current by an inverter. Notably, about 36% of mono-silicone and 55% of poly-silicone types as PV cell materials are broadly utilized to provide artificial light for poultry farm [31, 32]. This contributes to extending the day and improving the meat and egg production. Generally, solar PV module can either be roof installed or ground mounted for chicken houses [33–35]. Specifically, the Allen Family Foods Inc. mounted a 42 kW PV array with 314 m^2 area in the USA [31]. This PV array could output approximately 56112 kWh/year electrical energy, which could save about 78% energy consumption and £5700 operating expense per annum. Similarly, a 50 kW ground mount and a 49.82 kW roof mount solar PV arrays are utilized at the Cramble cross poultry farm. It is demonstrated that the power energy saving could achieve about 85% per annum compared to conventional heating system. The S.A. & D.E. poultry farm in the UK installed a 50 kW solar PV array, and found that the electrical energy production by the PV array could reach 42200 kWh/year, resulting in a 75% energy savings, £10854 operating cost savings and 11.03 tons CO_2 emission reduction per annum [36]. Additionally, in hot areas, the solar PV pumping system is needed for providing livestock watering usage and energy storage in the form of water in a water reservoir [37].

2.1.2 Solar thermal collector

Another vital factor for chicken house in some fields is heat to decrease the mortality rate of chicken [38]. Heat gains and losses from chicken and other resources are the core issue for the chicken house. Gad et al. [35] designed a flat plate solar thermal collector module to evaluate the system thermal efficiency and poultry production. The system composes of 12 horizontal copper tubes with 7 mm diameter which are embedded at the absorber surface plate. It is found that the system thermal efficiency is about 71.6% which can fulfill the poultry thermal demand. Brewer et al. [39] designed and installed three solar thermal collector modules to investigate the feasibility of heating poultry house. To be more specific, each module consists of twelve double glazed, copper plate collectors with 65 m^2 area, which is mounted on the roof with an angle of 45°. It is found that about 100% thermal energy demand can be fulfilled and about 75% energy consumption can be saved compared to traditional heating system.

2.1.3 Solar photovoltaic/thermal

Solar photovoltaic/thermal (PV/T) module can simultaneously produce heat and electricity for poultry farming by fully using the solar radiation lies in the overall solar spectrum ranging from 0.2–3 μm [40, 41]. Normally, the flat plate PV/T module is the most common category because of its constructional simplicity and building integration easiness [42, 43]. Meanwhile, it can shorten the payback time compared to the traditional PV module. Cui et al. [44] developed and installed an innovative PV/T integrated with ground copper pipe array system to decrease energy consumption and CO_2 emission for a poultry house in Newark, UK. It is revealed that the electricity and thermal output of the hybrid system could reach 11867 kWh and 30747 kWh, respectively. This contributes to obtaining about 70% electricity savings and 40% gas savings per annum, resulting in 6.23 tons for electricity and 5.65 tons for gas CO_2 emissions reduction.

2.2 Ventilation and wind turbine

The health level of the chicks is largely reliant on the indoor environmental temperature of the chicken shed. In winter, the indoor air temperature for broiler houses should be maintained ranging from 21 to 32°C, whereas the overheating and heat stress issues should be avoided in summer [45, 46]. Meanwhile, the ammonia (NH_3) and CO_2 are the two main harmful gases that must be controlled below their corresponding critical concentration levels of ~25 ppm and ~ 2500 ppm, respectively [47]. They can be removed from the poultry house by the ventilation fans. It is reported the energy consumption for the ventilation and cooling of a chicken shed can consume about 39.5% of the entire power energy usage and this value rises by 43.7% in laying hens [48, 49]. Specifically, Fawaz et al. [47] developed a parabolic concentrator solar thermal assisted with localized ventilation system for chicken brooding in Lebanon, and confirmed that the system is able to overlap 84% of thermal energy demand resulting in about 74% of energy savings and obtain a payback period of less than 5-year. Additionally, the wind turbine technology can be utilized to produce the power for the poultry farm. The energy could be generated by the force of the wind, and thus move the mills that are connected to electricity generators [50]. Small wind turbine, ranging from 0.4–40 kW, can fulfill the electrical demand of a whole poultry farm [50, 51]. Du et al. [51] designed and test a small-scale ventilation integrated with wind turbine energy system for a poultry house in China, and concluded that the electrical energy could be steadily produced 270 W/year by the wind turbine system, meanwhile, approximately 2074kWh/year renewable energy could be obtained, making up around 10.2% of entire energy demand and leading to 3.01 tons/year GHG emission reduction. Kapica et al. [52] developed a hybrid solar-wind turbine system for a chicken shed in the Central Europe. It is obtained that the system is conducive to a CO_2 emission reduction ranging from 0.11 to 0.22 kg/per kg in comparison with typical heating system. In addition to these typical cases, some other solutions have also been investigated involving the insulation thickness of chicken shed wall [53] and passive cooling system [54] to boost energy efficiency and save energy consumption.

2.3 Heat pump technology

Heat pump technology is commonly utilized in chicken house due to its advantage that can attain heating from one source including air/water/ground, and rise it to a suitable temperature for space water and heating. In the meantime, heat pump is one of the optimum solutions to heat shed in heating season with a fast return on investment [55].

2.3.1 Ground source heat pump

A ground source heat pump (GSHP) is able to extract heat source from soil for heating the chicken shed to obtain the optimum production performance, which is conducive to guaranteeing a desired indoor temperature for poultry shed and decreasing fuel requirement and CO_2 emissions. To be more specific, Choi et al. [56] developed a GSHP system utilized to provide the heating for a poultry house in Korea. It is concluded that the GSHP system could decline the operating cost since it could obtain more heating from ground for heating the shed. About 82% energy consumption could be saved in comparison to the conventional poultry shed during the operation period. Moreover, the concentration of the CO_2 in the GSHP poultry house is reduced by 2150 ppm compared to the traditional shed. Furthermore, the system COP is in the range from 3.5 to and 4.2, and the payback period is expected to be about 8-year. Kharseh and Nordell [57] built a GSHP integrated of the solar energy system to provide the thermal and electrical energy requirements for a chicken shed in Syria. It is indicated that the chicken weight is enhanced by about 5% ~ 6.8% on average. What is more, the energy consumption could be declined by 57% and has a payback time of about 6-year.

2.3.2 Water source heat pump

Water source heat pump (WSHP) is an effective approach of supplying water and space heating, meantime, it can be adapted to a range of bodies of water including canal, lake and river. The S.A. & D.E. Dixon chicken farm installed a WSHP system for rearing poultry house as a desired heat source for a heat pump operation in the UK [58]. The surface temperature of the lake is 4°C on average all over a year. It is obtained that about 33% of energy cost could be saved, which contributes to creating a 22% return on initial investment with a payback period of 5-year.

2.3.3 Air source heat pump

ASHP technology is the most common used in the chicken shed, which could extract heat from ambient air in heating season, meanwhile, it could reject heat outside in cooling season. Generally, the efficient of the ASHP technology is in the range from 300 to 400% [59]. For example, stonehouse farm installed six 14 kW ASHP units to supply the required heating to sustain the indoor temperature for two poultry sheds with a capacity of 20000 birds in the UK [60]. It is demonstrated that the energy consumption of each chick could reach below 1kWh, and system COP is in the range from 3 to 4. What is more, the economic benefit is approximately £11250 per annum with a payback period of 4.5-year.

2.4 Thermal energy storage technology

Thermal energy storage (TES) technology is typically considered for not only alleviating thermal demand of chicken shed but also stabilizing the indoor temperature variation for broilers growth. Hence, it is very important for keeping the stability of the temperature and relative humidity within poultry shed.

2.4.1 TES with phase change material

Phase change material (PCM) is usually regarded as a heat storage materials owing to its high thermal storage capacity that typically depends on latent heat in phase transition process. Additionally, PCM could absorb latent heat at a constant temperature, which is also the key factor for their application in TES with PCM system for poultry house. Zanaty [61] installed a thermal storage tank with PCM system to explored the influence of PCM latent heat on indoor environmental condition inside a chicken shed, and found that the replacement of the thermal tank from sensible water storage tank to latent heat water with embedded PCM results in a reduction of around 2% of the auxiliary energy demands. Moreover, there is a payback time of about 4-year in this case.

2.4.2 Trombe wall

A classical Trombe wall is used to absorb solar radiation for providing the indoor space water and heating, in the meantime, the external surface of the wall is colored black in order to increase the absorption rate. Additionally, the Trombe wall is usually mounted facing south to collect the solar thermal energy [62]. Currently, two categories of Trombe wall are usually adopted for the poultry shed because of low heat losses, high energy efficiency and convenient installation. Okonkwo and Akubuo [63] designed and built a Trombe wall heating unit to control the thermal energy requirement for the shed in Nigeria. Results concluded that the exterior surface temperature of the Trombe wall could achieve the maximum values of 60°C and the minimum values of 25°C, respectively, whereas the internal surface temperatures could reach ranging from 29.73 to 39.1°C. This indicates that this Trombe wall technology could accumulate and store adequate heat for the chicken shed in the tropics environmental condition.

3. Key findings and recommendations for future development

The development and usage of renewable and sustainable energy technologies are significant challenging research fields of poultry rearing. These technologies are advanced in power and heat production, fuel and gas savings, and the reduction of operating expense and GHG emission for poultry house. Meanwhile, they are conducive to sustaining the required indoor environmental conditions. Current researches have been identified obviously that the demand of the electrical and thermal energy could be fulfilled about 20–30% and 80–90% compared to the conventional heating system applied in poultry house, which therefore decreases fossil fuel consumption and GHG emission. Currently, the two renewable sources of energy production are wind and solar energy, both of which are seeing growing popularity for poultry healthy and high production. Specifically, wind power can generate huge electricity to meet the demand of poultry house, and wind turbine has a very low influence on the chick growing and environment. However, wind turbines need to be performed annual mechanical checks and servicing. In comparison, solar energy technology can be utilized for providing continuous heating, cooling, ventilation and lighting in poultry house, which requires less maintenance compared to other renewable energy technologies and supplies a potential long-term alternative available to anyone with a rooftop. Nevertheless, it cannot produce much electricity and heat energy when it is dark or cloudy, and there is a high installation costs. Despite these small issues, solar and wind energy are still the most preferable renewable energy sources for poultry house in the world. This is because they are the cheapest, fastest-growing, most reliable and do much less damage to nature and wildlife surrounding their sites as opposed to fossil fuels. In spite of being a vanguard and promising techniques, some perspectives and challenges for future technique investigation applied into the poultry house are put forward to formulate the framework for the future research interests as follows:

- Most existing technologies are relatively single, hence, more works should spend on the development of hybrid technologies. This contributes to enhancing the whole system performance and obtaining better production for the poultry farming.

- More concentration should be paid on the innovative solutions to break through the obstacles thus accelerate the practical application of these advanced technologies, instead of the fundamental theoretical researches those have been by far well done. Furthermore, the future of technical development forcefully relies on the latest study progress with regard to solar energy, wind energy, geothermal energy, PCM technologies as well as their dual roles for heating and cooling services.

- The accurate numerical models should be developed via the computer software in view of the influences of air temperature variation, wind speed, relative humidity, heating and cooling loads, building U-value, and ventilation rate. The aim is to provide a valuable instrument for researchers and engineers that make use of it to evaluate design alternatives and retrofit measures for various sizes and categories of chicken sheds under different environmental conditions.

- Similar technologies should be demonstrated and utilized to fattening sheep, fattening cows and fattening pigs for different national and regional fields. This is conducive to boosting usage of energy, reducing GHG emission as well as saving operating expense for livestock farming.

4. Conclusions

In this chapter, the state-of-the-art of the renewable and sustainable energy technologies are retrospected to replace for the traditional heating, ventilation, and air conditioning system within the poultry house. This can help to decline energy demands, operating expense, GHG emission and enhance farmers' profits. Consequently, some crucial outcomes are exemplified as follows:

- Solar energy technology could achieve electrical cost saving ranging from 30 to 85% for the poultry rearing. Moreover, the service lifetime of the solar energy technologies is able to maintain approximately 25 years with low maintenance cost for the poultry house.

- Ventilation and wind turbine technology could produce about 2000 kWh/year electrical and fulfill the energy demand of the whole poultry shed, which results in about 3.0 tons/year GHG emission reduction.

- Energy cost and average chicks weight of the GSHP poultry shed could be saved about 92% and raised by around 6.8% in comparison to the normal poultry shed, and the payback period is less 5-year. By contrast, the initial investment of the WSHP system is less than that of the GSHP because it does not need amount of ground works, and the payback time is around 6-year. Additionally, the ASHP is comparatively easy to be mounted, and needs low maintenance cost but often regarded as noisy.

- TES with PCM is a new usage in chicken shed to maintain indoor environmental quality during night time or cloudy day. This technology could store amount of heat with merely slight temperature variation, improve the effectiveness of TES and sustain the thermal storage for a long-term period.

- To sum up, these renewable and sustainable energy technologies can help save up to 85% energy consumption compared to the conventional chicken shed, and have a payback time of 3–8 years.

Acknowledgements

The authors would like to acknowledge the support from Innovate UK Agri-Tech Catalyst for the Welfare Enhanced Living Conditions for Healthier Chickens (WelChic) project (131897).

Conflict of interest

The authors declare that they have no known competing financial interests or personal relationships that could have appeared to influence the work reported in this paper.

Author details

Yuanlong Cui[1*], Xuan Xue[2] and Saffa Riffat[1]

1 Department of Architecture and Built Environment, The University of Nottingham, Nottingham, United Kingdom

2 School of Pharmacy, The University of Nottingham, Nottingham, United Kingdom

*Address all correspondence to: ezzyc14@yahoo.com

References

[1] Perkins KM, Munguia N, Ellenbecker M, Moure-Eraso R, Velazquez L. COVID-19 pandemic lessons to facilitate future engagement in the global climate crisis. Journal of Cleaner Production 2020; 125178.

[2] Martorell-Marugán J et al. DatAC: A visual analytics platform to explore climate and air quality indicators associated with the COVID-19 pandemic in Spain. Science of The Total Environment 2021; 750: 141424.

[3] Barouki R et al. The COVID-19 pandemic and global environmental change: Emerging research needs. Environment International. 2021; 146: 106272.

[4] Mofijur M et al Impact of COVID-19 on the social, economic, environmental and energy domains: Lessons learnt from a global pandemic. Sustainable Production and Consumption 2021; 26: 343-359.

[5] Manzanedo RD, Manning P. COVID-19: Lessons for the climate change emergency. Science of The Total Environment 2020; 742: 140563.

[6] Wang Y, Sun X, Wang B, Liu X. Energy saving, GHG abatement and industrial growth in OECD countries: a green productivity approach. Energy 2020; 194:116833.

[7] World Meteorological Organization. Global Climate in 2015-2019. Available from: < https://public.wmo.int/en/media/press-release/global-climate-2015-2019-climate-change-accelerates> [Accessed: 2019-09-22].

[8] Wiedemann SG, McGahan EG, Murphy CM. Resource use and environmental impacts from Australian chicken meat production. Journal of Cleaner Production 2017; 140: 675-684.

[9] Pereira JLS. Assessment of ammonia and greenhouse gas emissions from broiler houses in Portugal. Atmospheric Pollution Research 2017; 8 (5): 949-955.

[10] Zareei S. Evaluation of biogas potential from livestock manures and rural wastes using GIS in Iran. Renewable Energy 2018; 118: 351-356.

[11] Kapica J, Pawlak H, Ścibisz M. Carbon dioxide emission reduction by heating poultry houses from renewable energy sources in Central Europe. Agricultural Systems 2015; 139: 238-249.

[12] Zanaty HE. A techno-economic study for heating poultry houses using renewable energy. Available from: <http://citeseerx.ist.psu.edu/viewdoc/download?doi=10.1.1.843.2738&rep=rep1&type=pdf> [Accessed 2015-11-02].

[13] Rojas-Downing MM, Nejadhashemi AP, Harrigan T, Woznicki SA. Climate change and livestock: Impacts, adaptation, and mitigation. Climate Risk Management. 2017; 16: 145-163.

[14] Huang D, Guo H. Diurnal and seasonal variations of greenhouse gas emissions from a commercial broiler barn and cage-layer barn in the Canadian Prairies. Environmental Pollution. 2019; 248: 726-735.

[15] Vitt R, Weber L, Zollitsch W, Hörtenhuber SJ, Hörtenhuber J, Niebuhr K, Piringer M. Modelled performance of energy saving air treatment devices to mitigate heat stress for confined livestock buildings in Central Europe, Biosystems Engineering 164 (2017) 85-97.

[16] Tamvakidis S, Firfiris VK, Martzopoulou A, Fragos VP, Kotsopoulos TA. Performance evaluation of a hybrid solar heating system for farrowing houses. Energy and Buildings 2015; 97: 162-174.

[17] Ghahramani A, Bowran D. Transformative and systemic climate change adaptations in mixed crop-livestock farming systems. Agricultural Systems. 2018; 164: 236-251.

[18] Sneessens I, Sauvée L, Randrianasolo-Rakotobe H, Ingrand S. A framework to assess the economic vulnerability of farming systems: Application to mixed crop-livestock systems. Agricultural Systems 2019; 176: 102658.

[19] Purdy A, Pathare PB, Wang Y, Roskilly AP, Huang Y. Towards sustainable farming: Feasibility study into energy recovery from bio-waste on a small-scale dairy farm. Journal of Cleaner Production 2018; 174: 899-904.

[20] Torshizi MV, Mighani AH. The application of solar energy in agricultural systems. Journal of Renewable Energy and Sustainable Development (RESD) 2017; 3 (2): 2356-8569.

[21] Ali SM, Dash N, Pradhan A. Role of renewable energy on agriculture. International Journal of Engineering Sciences & Emerging Technologies 2012; 4 (1): 51-57.

[22] Seyfi SU. Comparison of the energy efficiency of dairy production farms using different housing systems. Environmental Progress & Sustainable Energy 2012; 32: 1202-1208.

[23] Bell MJ, Cloy JM, Rees RM. The true extent of agriculture's contribution to national greenhouse gas emissions. Environment Science & Policy 2014; 39: 1-12.

[24] Ghahramani A, Bowran D. Transformative and systemic climate change adaptations in mixed crop-livestock farming systems. Agricultural Systems 2018; 164: 236-251.

[25] Sharif A, Meo MS, Chowdhury MAF, Sohag K. Role of solar energy in reducing ecological footprints: An empirical analysis. Journal of Cleaner Production 2021; 292: 126028.

[26] Behura AK, Kumar A, Rajak DK, Pruncu CI, Lamberti L. Towards better performances for a novel rooftop solar PV system. Solar Energy 2021; 216: 518-529.

[27] Das D, Kalita P, Roy O. Flat plate hybrid photovoltaic- thermal (PV/T) system: A review on design and development. Renewable and Sustainable Energy Reviews. 2018; 84: 111-130.

[28] Zhou B, Pei J, Nasir DM, Zhang J. A review on solar pavement and photovoltaic/thermal (PV/T) system. Transportation Research Part D: Transport and Environment 2021; 93: 102753.

[29] Sharaf OZ and Orhan MF. Concentrated photovoltaic thermal (CPVT) solar collector systems: part I – fundamentals, design considerations and current technologies. Renewable and Sustainable Energy Reviews. 2015; 50: 1500-1565.

[30] Zhou J, Zhang Z, Ke H. PV module temperature distribution with a novel segmented solar cell absorbance model. Renewable Energy 2019; 134: 1071-1080.

[31] First solar-powered poultry house set 2007. Available from: < http://www1.udel.edu/PR/UDaily/2007/dec/solar121806.html> [Accessed: 2007-02-04].

[32] Byrne J, Glover L. The potential of solar electric applications for Delaware's poultry farms, Center for Energy and Environmental Policy. University of Delaware, USA; 2005.

[33] Cramble cross (poultry) farm (Ground and Roof mounted). Available from: < https://duncanrenewables.co.uk/projects/view/63/cramble-cross-farm> [Accessed: 2013-06].

[34] Misbrener K. Southern solar system adds solar-plus-storage to poultry house for Auburn University experiment. Available from: < https://www.solarpowerworldonline.com/2019/06/tyson-foods-auburn-university-off-grid-solar-poultry-house/> [Accessed: 2019-06].

[35] Gad S, El-Shazly MA, Wasfy KI, Awny A. Utilization of solar energy and climate control systems for enhancing poultry houses productivity. Renewable Energy 2020; 154: 278-289.

[36] S.A. & D.E. Dixon Poultry Farm, Shropshire. Available from: < http://www.salopenergy.com/commercial/case-studies/sa-and-de-dixon-ltd-vownog-porth-y-waen-oswestry-shropshire/> [Accessed: 2015-07].

[37] Rota A. Livestock and renewable energy. Livestock Thematic Papers Tools for project design. Available from: <https://www.ifad.org/documents/38714170/39148759/Livestock+and+renewable+energy.pdf/61af921d-a886-4558-b2d1-b97a0650ce06> [Accessed: 2012-10].

[38] Hussein MS, Burra KG, Amano RS, Gupta AK. Effect of oxygen addition in steam gasification of chicken manure. Fuel. 2017; 189: 428-435.

[39] Brewer RN, Flood CA, Koon JL. Solar Heating of Poultry Houses: 3. Design and Testing of Research Facilities. Poultry Science 1981; 60(8): 1802-1806.

[40] Mukunda N, Praveen KK, Sachin MV. Role of renewable energy on agriculture. International Journal of Scientific & Engineering Research 2018; 9 (7): 2229-5518.

[41] Al-Waeli AHA, Sopian K, Kazem HA, Chaichan MT. Photovoltaic thermal PV/T systems: A review. International Journal of Computation and Applied Sciences 2017; 2 (2): 62-67.

[42] Cui Y, Zhu J, Zoras S, Zhang J. Comprehensive review of the recent advances in PV/T system with loop-pipe configuration and nanofluid. Renewable and Sustainable Energy Reviews 2021; 135: 110254.

[43] Das D, Kalita P, Roy O. Flat plate hybrid photovoltaic-thermal (PV/T) system: A review on design and development. Renewable and Sustainable Energy Reviews 2018; 84: 111-130.

[44] Cui Y, Elmer T, Gular T, Su Y, Riffat S. Feasibility of hybrid renewable heating system application in poultry house: a case study of East Midlands, UK. International Journal of Low-Carbon Technologies 2020; 00: 1-16.

[45] ASHRAE, 2011. Handbook-Heating, Ventilating and Air-Conditioning Applications (I-P Edition). American Society of Heating, Refrigerating and Air-Conditioning Engineers.

[46] Zhao Y, Shepherd TA, Li H, Xin H. Environmental assessment of three egg production systems–Part I: monitoring system and indoor air quality. Poultry Science 2015; 94: 518-533.

[47] Fawaza H, Abiad MG, Ghaddar N, Ghali K. Solar-assisted localized ventilation system for poultry brooding. Energy and Buildings 2014; 71: 42-154.

[48] Costantino A, Fabrizio E, Biglia A, Cornable P, Battaglini L. Energy use for climate control of animal houses: the state of the art in Europe. Energy Procedia 2016; 101: 184-191.

[49] Costantino A, Fabrizio E, Ghiggini A, Bariani M. Climate control in broiler houses: A thermal model for the calculation of the energy use and indoor environmental conditions. Energy and Building 2018; 169: 110-126.

[50] Hong SW, Lee IB, Seo IH, Kwon KS. The design and testing of a small-scale

wind turbine fitted to the ventilation fan for a livestock building. Computers and Electronics in Agriculture 2013; 99: 65-76.

[51] Du L, Hu C, Yang c, Yang L, Du H, Li Q, Yu C, Xie L, Jiang X. Investigation of a preliminary ventilation energy-recovery system for poultry houses. Computers and Electronics in Agriculture 2020; 175: 105521.

[52] Kapica J, Pawlak H, Ścibisz M. Carbon dioxide emission reduction by heating poultry houses from renewable energy sources in Central Europe. Agricultural Systems 2015; 139: 238-249.

[53] Kucuktopcu E, Cemek B. A study on environmental impact of insulation thickness of poultry building walls. Energy 2018; 150, 583-590.

[54] Firfiris VK, Martzopoulou AG, Kotsopoulos TA. Passive cooling systems in livestock buildings towards energy saving: A critical review. Energy and Building 2019; 202: 109368.

[55] Heat pumps for pig- and poultry houses. Available from: < https://inno-plussystems.com/en/products/heatpump/> [Accessed: 2014-04].

[56] Choi HC, Salim HM, Akter N, Na JC, Kang HK, Kim MJ. Effect of heating system using a geothermal heat pump on the production performance and housing environment of broiler chickens. Poultry Science. 2012; 91: 275-228.

[57] Kharseh M, Nordell B. Sustainable heating and cooling systems for agriculture. International Journal of Energy Research. 2011; 35: 415-422.

[58] Davies J. Carp lake used to heat poultry sheds. 2014 Available from: < https://www.fwi.co.uk/livestock/poultry/broilers/carp-lake-used-to-heat-poultry-sheds> [Accessed: 2014-04].

[59] Davies J. Air-source heat pumps power poultry shed. Available from: < https://www.fwi.co.uk/business/air-source-heat-pumps-power-poultry-shed> [Accessed: 2014-05].

[60] Stonehouse poultry farm. Available from: <https://library.mitsubishielectric.co.uk/pdf/download_full/1051> [Accessed: 2014-07].

[61] Zanaty HE. A techno-economic study for heating poultry houses using renewable energy. The American University in Cairo. 2015. Available from: <http://dar.aucegypt.edu/bitstream/handle/10526/4281/MSc%20Thesis%20Report_HZ.pdf?sequence=1> [Accessed: 2015-10-02].

[62] Hu Z, He W, Ji J, Zhang S. A review on the application of Trombe wall system in buildings. Renewable and Sustainable Energy Reviews. 2017; 70: 976-987.

[63] Okonkwo WI, Akubuo CO. Trombe wall system for poultry brooding. International Journal of Poultry Science. 2007; 6: 125-130.

Chapter 3

Poultry Meat Production in the South American Andes

Manuel E. Paredes Arana

Abstract

This chapter will explain some of the research carried out in the production of poultry meat in natural hypobaric areas, where the development of industrial poultry farming is not traditional. Relevant aspect of the production of chickens, hens and turkeys for meat purposes will be clarified, as well as their benefits, and characteristics and why it must still be carried out in the Peruvian Andes. Physiological aspects of birds, use of unconventional food; as well as the productive evaluation of poultry species not used intensively, are approached with the purpose of generating and stimulating the obtaining of meat as an economic source for the rural sector and small companies.

Keywords: turkey meat, indigenous chicken, ecological poultry, lupine, hypobaric conditions

1. Introduction

In South America and Peru, poultry production is developed industrially in the coastal region with geographical heights that guarantee conditions of oxygen concentrations necessary for the growth of improved birds. However, it is necessary to take advantage of very important resources such as land, water and human resources in Andean areas, under hypobaric conditions; being necessary to demonstrate that the raising of birds and the production of meat is possible in such conditions.

For this reason, a series of works have been developed that guarantee and corroborate these statements. In Lavras, MG, Brazil, located at 919 m altitude, selenomethionine improved the weight gain and feed conversion in broilers of 1 to 42 d [1]. Broilers are raised to cool conditions at high altitudes in Shahrekord, Iran, at 2,100 m above sea level, with good cardiac parameters due use flavonoids bioactive compounds [2] and with dietary arginine supplementation to boost performance [3]. In Cajamarca, Peru at 2,700 m above sea level, the commercial turkey is raised up to fourteen weeks old with body weights 8 and 12 kg in females and males [4]. This review shows some works carried out in the Peruvian Andes region, with emphasis on the production of chicken and turkey.

2. World and regional poultry production

The first producer of chicken meat in the world has been the United States (17.6%), followed by Brazil (12.5%), then China (12.4%), Russia (4.1%), India

(3.2%) and Mexico (2.9%) mainly. In this ranking, Peru is ranked number 18. Brazil is indisputably the largest chicken producer in Latin America with 6,468.6 million chickens slaughtered in 2019. Brazilians represent 51.6% of all chicken production at the regional level, out of a total of more than 12,500 million birds. The 10 largest chicken production companies in the region are located in Brazil, Mexico, Peru, Colombia, Argentina and Central America. In 2019, Peru was the fourth largest producer of chickens in Latin America and the fifth largest producer of eggs in the region.

Peru, during the years 2017, 2018 and 2019, occupies the first position of per capita consumption of chicken in Latin America, with a consumption of 51.1 kilos of chicken per person in 2019. Behind is Argentina (46.6 kilos), Bolivia (43 kilos), Brazil (42.6 kilos) and Panama (41.28 kilos). The average consumption of the region was 31.42 kilos.

In 2019, at least 28.24 million turkeys were produced in Latin America, with Brazil being the largest producer of this animal protein with 13 million turkeys or 46% of the total poultry produced in the region. In terms of the countries that produce the most turkeys, behind Brazil is Peru (5.43 million turkeys), Chile (6.05 million), Ecuador (1.33 million), Mexico (1.18 million) and Bolivia and Colombia (each with 400,000 turkeys). produced in 2019).

3. Dietary inclusion of alfalfa meal in organic type broiler chicken

In Peru, 31 tons of meat from laying hens that have finished their productive cycle and about 46 tons of meat from backyard poultry and commercial hybrid breeding hens are consumed annually [5]. These birds are of advanced age and the laying hens at the end of the laying cycle end up with bone and liver problems [6], have low muscle mass and low carcass performance; Although it has been determined that meat from commercial laying hens, under the same diet, due to its composition in fatty acids, is more favorable for human health than meat from local breeds [7]; however, its mature and cross-linked thickened muscle fibers make it lose tenderness [8].

On the other hand, hedonism, health and the economy are factors that strongly influence the consumer [9] and drive the intake of organic foods [10]; Therefore, production systems with principles based on animal welfare and organic production with good sensorial quality of meat are increasing [11–13]. Also, according to the European Commission, 2008, organic meat can be obtained from fast-growing broiler chicken, fed diets that include green, dried or preserved forage, slaughtered after 81 days of age, without specifying the amount of pasture that the birds would have to eat. Several rearing systems have been evaluated [14], including the duration of the free rearing period, reporting that the sources of dietary fiber affect the oxidative stability of the meat, the blood parameters, and the content of proteins, cholesterol, fatty acids and triglycerides [15], as well as an increase in the weight of the gizzard and alterations in growth [16]. The antioxidant bioactive compounds of alfalfa, Phleum sp., *Dactylis glomerata*, Italian ryegrass (*Lolium multiflorum*) and balansa clover (*Trifolium michelianum*), among other forages, have been shown to improve the content of polyunsaturated fatty acids (PUFAs) n-3 and the proportion between PUFA n-6 and n-3 [17]. It has been determined that the saponins of dehydrated alfalfa have a hypocholesterolemic effect in chicken meat [18] and that the natural extract of alfalfa reduces the deposition of abdominal fat and improves the antibody titter against Newcastle disease, without an adverse effect on bird performance [19]. Additionally, alfalfa xanthophyll esters are known to produce better skin pigmentation in broilers [20].

As alfalfa is a widely available forage in Cajamarca, it was included in dehydrated form in the diet of the fast-growing chicken aimed at achieving the classification of organic chicken, with the hypothesis that alfalfa has beneficial effects on growth, carcass weight, weight digestive organs and lymphoid organ weight as it is an immunomodulatory food [21, 22]. For this reason, the effects of three levels of alfalfa meal (AM) included in the diet of female chicken, Hubbard line, supplied from 35 to 84 days of age [23].

The data suggest that the variables feed consumption and feed efficiency were not affected by the inclusion of dehydrated forage in the diet. Birds that consumed diets with 10% AM had lower carcass weight; however, no such differences are observed in carcass performance, indicating that the difference in carcass weight is due to the size of the live bird. On the other hand, the absolute and relative abdominal fat weights were higher with the HA-containing diets. The greater accumulation of abdominal fat in chickens agrees with some researchers [24] who by increasing the crude fiber of the diet from 3.19 to 3.52% in guinea fowl increased abdominal fat, without finding different carcass yields. Carcass performance is associated with the FC content of the diet; where lower feed intake and pasture consumption can have a negative effect on the carcass performance of birds compared to birds without access to pastures. On the other hand, the greater accumulation of abdominal fat in birds that consumed AM may be due to the fact that alfalfa strengthens the immune system, and this regulates lipid metabolism, hence, as there were no health challenges, lipolysis did not occur, but increased of abdominal fat deposition.

Not observed effects of AM on the absolute and relative weights of the digestive organs on the absolute value of the weight of the organs of the gastrointestinal tract, except for the intestines. In relative terms, the differences observed in the total weight of the GIT and higher gizzard development were also reflected in favor of the birds with 10% AM. According to the fiber content in the diets of the present study, from 3.56% (0% AM) to 4.54% (5% AM) produced an increase of 5.3 and 3% in the relative weight of intestines and gizzard, respectively. Likewise, the relative weight of intestines and gizzard increased to 19% with 5.56% fiber (10% AM), which could be due to the fact that dietary fiber intake causes an increase in the gastrointestinal tract of the birds, as a result of the increase in size. of the blind [25]. Similar results have been reported [26] with an 8.7% increase in gizzard weight of 35-day-old birds with the inclusion of 4% straw in the diet, and also when including 3% sunflower husk in the diet of the bird. Broilers from 0 to 21 days of age, resulting in an increase in gizzard weight [27]. Although it can be concluded that the inclusion of 10% AM in the diet in the present study produces a greater development of intestines and gizzards; It should be considered that the size of the gizzard can not only be increased by the inclusion of fibrous food in the diet, but also when the food particle is large [28].

4. Growth and carcass characteristics of six genotypes in chickens reared in Andean region of northern Peruvian

The per capita consumption of chicken meat in Peru exceeds 51 kg, with imports exceeding 34 and 13 million tons of frozen chicken and hen meat, respectively [29], showing there is still demand to be satisfied. In this regard, statistics indicate that poultry meat is the main driver of the meat sector in the world, due to the need for animal protein and because it is cheaper than red meat, with poultry meat being preferred by producers and consumers in developing countries [30].

The poultry meat supply comes mainly from fast-growing broiler genetic lines with high carcass yields; although, its flavor characteristics and meat qualities are not highly appreciated by some consumers [31]. There is also the sector of native or creole poultry meat, eaten locally due to the consumer' perception that highlights its good taste and organic characteristics, without negative impact on human health as it is raised free of antibiotics [32], whose meat is low in fat and cholesterol [33], and its production is complemented with eggs rich in protein and iron [34].

Faced with such a situation, there is a supply of different genotypes of birds with colored plumage, by incubator companies registered in the Peruvian poultry system; reporting an approximate placement of 15 million crossed BB chickens nationwide, which is why the growth of this sector has been accentuating unconventional production with the use of the intensive rearing system at the level of small farms, without even having more technical information about the productive performance of these birds and especially in the Andes region.

Researchers from several countries are studying the native chicken, having determined in five Egyptian chicken breeds a genetic difference between pure native and improved native [35]. In high mountain areas of Thailand, they characterized different phenotypes of indigenous chicken reared in various production systems [36]. In Saudi Arabia they used microsatellite markers to assess the genetic diversity of their native chicken populations [37]. In Indonesia they did similar studies with single nucleotide polymorphism markers [38]. In northern Thailand they compared the meat of indigenous chickens and crossbred chickens of imported origin, in southern Thailand they determined the heritability and genetic correlations between growth indicators and meat quality of a line of improved indigenous chicken [39]. Chee chicken from Thailand was crossed with specialized lines of meat and other native breeds in order to improve meat quality and carcass performance [40]; while in India they conducted studies to conserve and characterize the germplasm of the Aseel chicken, which is considered in danger of extinction [41]; having been compared to the Kadaknath breed in terms of growth, egg production, semen quality and health indicators. In South Africa, indigenous chicken was fed canola meal instead of soybean without negative impact on carcass, organ size and meat quality [42]. Determined the proximal composition and amino acid content of the meat of the 180-day-old Italian chicken Poverara [43]; also, in Italy evaluated the growth of the local races Berlanda and Padovana [44] and work on the conservation of germplasm of other races such as: Ermellinata di Rovigo, Pépoi, Robusta Lionata and Robusta Maculata [45]. The Spaniards evaluated the crossing of the indigenous Galician rooster Mos with the line chicken Sasso T-44 [46]. In Ghana they carried out hematobiochemical studies in indigenous chickens to determine better performance and adaptability of the various genotypes [47]. Therefore, in Peru, among a variety of studies to be carried out on Creole chickens, it is necessary to evaluate the productive performance of the available birds of local origin and recently introduced, determining their biological capacity to thrive in high areas of the Andes.

With this objective, six chicken genotypes were comparatively evaluated: Native French, Hubbard red, Improved Peruvian Creole, Pure Peruvian Creole, Babcock Brown male chickens that are of the laying genetic line, contrasted with the Hubbard white, specialized in meat production; in order to determine its growth parameters, carcass characteristics, weight of internal organs and proximal composition of meat in conditions of the Cajamarca valley at 2684 meters above sea level.

The performance of the growing chickens evaluated up to 13 weeks of age indicated que both varieties of Hubbard chicken showed higher growth speed, followed by French Native and Improved Creole, as opposed to improved Peruvian Creole and Babcock Brown chickens that showed the lowest rates of increase in body mass. Proximal composition of the six types of chicken breast meat differ markedly in fat. Breast meat from the Creole, Native French and Babcock line breeds has the lowest content fat, while the breast meat from Hubbard broilers has the highest amount. There is a trend for higher protein in the meat of native French and Creoles compared to Hubbard chickens, possibly due to the fact that lipidosis involves the replacement of muscle fibers by adipose tissue [48].

5. Lupine seed meal (Lupinus mutabilis) in Turkey feed

The interest in the food sector to identify, produce and use new protein sources for the nutrition of humans and animals is a task that has no end [49]. The national and global poultry sector is looking for alternatives in this field in order to include ingredients that efficiently replace or exceed soybean cake (TS), the most widely used protein ingredient in poultry feed, as many countries, such as Peru, they do not have an industrial production of this oil seed, depending on its permanent importation. However, in the Peruvian highlands there are various protein crops of the legume family, grown locally, such as chocho, commonly known as lupine, tarwi or lupine. Studies with yellow lupine (*Lupinus luteos*) seeds indicate that it could be included in the diet of broilers up to a proportion of 20 g/kg [50].

The main limiting factor for the use of lupine seeds yellow in the diet of poultry is the high concentration of non-starch polysaccharides (NAP), which reduce the nutritional value of the seeds [51], where raffinose can hinder the transport of nutrients through the intestinal wall [52]. The protein value of the *Lupinus mutabilis* (LM) seed harvested in the Peruvian highlands is similar to that of soybean, but its use has been limited by another additional factor, such as the presence of large amounts of quinolizidine alkaloid structures, with certain degree of toxicity and strongly bitter taste [53].

Other species and varieties of lupines such as yellow lupine, white lupine (*Lupinus albus*) or blue lupine (*Lupinus angustifolius*) have been included in the finisher diets of 18-week-old turkeys at levels of up to 180 g / kg, without affect gastrointestinal function and without adverse effects on the growth or quality of the meat [54], even though it can cause an increase in the weight of the gizzard, a decrease in the pH of the contents of the gizzard and problems in viscosity of the small intestine digest [55]. On the other hand, Samulikowska et al. [56] indicated that starter diets with 100 g/kg of yellow lupine seeds reduced body weight in broilers due to a lower feed intake, in addition to changes in the gastro-tract intestinal caused by increased concentrations of PNA and raffinose. However, inclusion of LM in feeding turkeys in their first weeks of life is not documented. Likewise, LM contains more than 20 types of alkaloids or toxic substances that could cause hemolysis and anemia [57], as happens with other plant alkaloids whose toxic effect can be combined with the ferric ion in the blood, affecting the transport capacity of erythrocyte oxygen; as has been demonstrated in young fish and ducks [58].

The effect of four dietary levels of lupine seed meal (*Lupinus mutabilis*, HCH) on growth performance, intestinal tract development and hematological values were evaluated in male turkeys of the one-day BUT line [59]. The turkeys were distributed in four groups with different diets for 56 days: L0 (control) diet without HCH, and L30, L60 and L90 diets containing 30, 60 and 90 g/kg of HCH, respectively. Increased levels of HCH inclusion in the diet, added at the expense of soybean meal contributed to an increase in fiber concentrations in the experimental diets. Daily feed intake decreased in HCH treatments and feed conversion and bodyweight gain of birds were affected in treatments of 60 and 90 g/kg of HCH inclusion. High dietary levels of HCH led to an increase of the digesta in the ileum and in the cecum, an increase in the relative tissue mass, as well as a decrease in pH values of the digesta. The increase in HCH in the diet did not alter the hematological values. It is concluded that HCH can be considered as a safe and effective alternative for dietary inclusion at levels of 30 g/kg in starting diets for young turkeys.

6. Vitamin C and the productive performance of Turkey reared in a natural hypobaric environment

The Andean zone of Peru (natural hypobaric environment) has more than eight million of inhabitants [60] grouped in more than a million families; A lot of they consume turkey meat at Christmas parties, maintaining the custom of consuming it from birds raised in the same area and from native species; However, in the last few years the supply of turkeys from lines commercial genetics has increased and the inhabitants of the Andean cities have reversed their preferences for this type of turkey due to its lower cost and higher meat yield, but it relates the breeding turkey in the same area with the good taste, which would have a biochemical basis because the transport of the turkey during several hours from the farm to downtown processing produces high levels of corticosterone [61], in addition to bruising that the bird suffers during transfer; so the consumer prefers to buy live birds, which they slaughter after a previous inspection of its appearance bodily.

There are three important factors that hinder optimal rearing of the turkey for fattening: genetics, cold weather and lower oxygenation; in this regard, the intense selection of broilers for rapid growth and high meat yield is a great achievement, but it has led to an increase in the incidence of metabolic syndromes caused by the hypoxemic condition resulting from an imbalance between the requirement and the oxygen supply [62]. These birds in conditions of temperatures low and high geographical areas have problems of pulmonary hypertension and ascites [63]. Birds raised to high altitude achieve higher body weights lower than their level zone counterparts from the sea [64, 65], where there is a minimal incidence of ascites and right ventricular hypertrophy, lower red blood cell counts and low hemoglobin values [66]. Also, fast-growing broilers present dilated cardiomyopathies without presenting ascites, these cases being differentiated by elasticity and density of the arterial vascular structure, and the thickness of the fibers in the vena cava. These cases of cardiomyopathies with or without the associated presence of ascites occur very often in turkeys fattening raised in the Peruvian highlands. The low ambient temperature induces appearance of ascitic syndrome due to increased metabolic rate and oxygen requirement. In these conditions of environmental stress are produce at the mitochondrial level species of reactive oxygen such as oxygen ions, free radicals and peroxides that cause changes in the shape, structure and function of this key organelle in metabolism, causing oxidative stress [67]. Vitamin C (VC) is known to be an antioxidant capable of eliminating reactive oxygen species induced by low ambient temperatures, reducing stress [68–70], which in turkeys reflected by the increase in the heterophile/lymphocyte ratio [71]. VC prevents pulmonary vascular remodeling that causes ascites syndrome in broilers [72]. VC supplementation restores xanthine oxidase activity that meets the function of buffering the effects of superoxide in broilers raised under hypobaric hypoxia [73]. Likewise, CV is a powerful activator of the immune system, enhancing the antioxidant activity of other dietary compounds and improving the productive performance of growing species by reducing stress in environments exposed to ammonia production.

Also, it should be noted that for the formulation and production of animal feed turkey for fattening is not recommended to include VC because it is synthesized from glucose, via glucuronic acid and lactone from gulonic acid in the presence of the enzyme L gulonolactone oxidase; unlike fat soluble vitamins and B complex vitamins that must be included in specific amounts in the diet of the turkey using vitamin premixes; therefore, the supplementation of VC in the diet could generate in the turkey for fattening subject to permanent selection and raised under environmental conditions hypoxia, better productive results in terms of growth, reduction of stress and immune response against the main diseases that occur in this species.

With the objective of determining the best level of dietary supplementation of vitamin C in the fattening turkey raised in a natural hypoxic environment of the Cajamarca valley an experiment was carried out. Vitamin C decreased heterophile/lymphocyte levels. Compared to the control group, vitamin C at a level greater than or equal to 1000 mg/kg decreased bird mortality, reduced the stress indicator, increased the antibody titer, but did not improve growth parameters. Consequently, dietary supplementation with vitamin C could regulate productive performance, by reducing the level of mortality, promoting immune function, and improving the state of stress in commercial turkeys, reared in a natural hypoxic environment. These results encourage commercial turkey meat production in hypoxic areas.

7. Opportunities and challenges to produce poultry in South America

World meat production in 2016 increased only 1%. Among the various sectors, poultry and beef production increased. The increase in world exports of meat products was led by Brazil and the European Union, followed by the United States, and sales also increased in South American countries such as Argentina and Paraguay.

Global meat production is projected to be 13% higher in 2026 than in the 2014–2016 base period. Developing countries are estimated to account for the vast majority of the total increase, given the more intensive use of food in the production process and the lower cost of human resources. Poultry meat is the main driver of growth in total meat production, due to the greater world demand for this animal protein, which is cheaper than red meat. Low production costs and lower product prices contribute to poultry meat being the preferred meat for both producers and consumers in South American countries.

8. Conclusion

The physiology of birds for meat purposes is adaptable to hypobaric challenges, with convenients feeding strategies and programs, to generate in birds an efficient use of feed nutrients, which synthesized as poultry body nutrients will be very necessary to human nutrition in of highlands.

The production of fast-growing poultry meat in territories with natural hypobaric conditions is technically feasible. You can use food produced in the same region. Organic chickens, indigenous chickens, turkeys and muscovy ducks can be raised on a small scale and industrially.

Conflict of interest

I have no conflict of interest'.

Author details

Manuel E. Paredes Arana
Faculty of Engineering in Livestock Sciences, National University of Cajamarca, Cajamarca, Perú

*Address all correspondence to: mparedes@unc.edu.pe

References

[1] Silva VA, Clemente AHS, Nogueira BRF, de Carvalho AC, Boas de Freitas LFV, de Lemos Souza Ramos A, Bertechini AG. Supplementation of selenomethionine at different ages and levels on meat quality, tissue deposition, and selenium retention in broiler chickens. Poultry Science. 2019; 98:2150-2159. http://dx.doi.org/10.3382/ps/pey569

[2] Ahmadipour B, Kalantar M, Schreurs NM, Raza SHA, Khan R, Khan S, El-Aziz AHA, Memon S, Ullah I, Samira A. Flavonoid bioactive compounds of hawthorn extract can promote growth, regulate electrocardiogram waves, and improve cardiac parameters of pulmonary hypertensive chickens. Poultry Science. 2020; 99:974-980. https://doi.org/10.1016/j.psj.2019.10.022

[3] Khajali F, Tahmasebi M, Hassanpour H, Akbari MR, Qujeq D, Wideman RF. Effects of supplementation of canola meal-based diets with arginine on performance, plasma nitric oxide, and carcass characteristics of broiler chickens grown at high altitude. Poultry Science. 2011; 90 :2287-2294. DOI: 10.3382/ps.2011-01618

[4] Paredes M, Terrones A, Hosbán C, Ortiz P. Effect of dietary supplementation with vitamin C on the productive performance, stress and immune response of turkey reared in a natural hypoxic environment. Scientia Agropecuaria. 2020; 11(3): 357-364. DOI: 10.17268/sci.agropecu.2020.03.07

[5] [MINAGRI] Ministerio de Agricultura y Riego. Producción y comercialización de productos avícolas. Lima, Perú. Boletín estadístico mensual del sector avícola 2018. [Internet]. Available from: https://www.minagri.gob.pe/portal/boletin-estadistico-mensual-de-la-produccion y-comercializacion-avicola/sector-avicola-2018

[6] Molnár A, Zoons J, Buyse J Delezie E. Extending the laying cycle of laying hens. In: XVII European Symposium on the Quality of Eggs and Egg Products. 2017. Edinburgh, UK

[7] Rizzi C, Chiericato GM. Chemical composition of meat and egg yolk of hybrid and Italian breed hens reared using an organic production system. Poultry Science. 2010; 89:1239-1251. DOI: 10.3382/ps.2008-00045[22]

[8] Vaithiyanathan S, Naveena BM, Muthukumar M, Girish PS, Ramakrishna C, Sen AR, Babji Y. Biochemical and physicochemical changes in spent hen breast meat during postmortem aging. Poultry Science. 2008; 87: 180-186. DOI:10.3382/ps.2007-00068

[9] Brunsø K, Fjord TA, Grunert KG. Consumers' food choice and quality perception. MAPP Working Paper 77. 2002 Aarhus, Denmark: Aarhus School of Business. 60 p.

[10] Fanatico AC, Pillai PB, Cavitt LC, Emmert JL, Meullenet JF, Owens CM. Evaluation of slower growing broiler genotypes grown with and without outdoor access: sensory attributes. Poultry Science. 2006; 85: 337-343. DOI:10.1093/ps/85.2.337

[11] Rizzi C, Marangon A, Chiericato GM. Effect of genotype on slaughtering performance and meat physical and sensory characteristics of organic laying hens. Poultry Science. 2005; 86:128-135. DOI:10.1093/ps/86.1.128

[12] Ponte PI, Prates JA, Crespo JP, Crespo DG, Moura JL, Alves SP, Bessa RJ. Improving the lipid nutritive value of poultry meat through the

incorporation of a dehydrated leguminous-based forage in the diet for broiler chicks. Poultry Science. 2008; 87: 1587-1594. DOI:10.3382/ps.2007-00446

[13] Tong HB, Wang Q, Lu J, Zou JM, Chang LL, Fu SY. Effect of free-range days on a local chicken breed: growth performance, carcass yield, meat quality and lymphoid organ index. Poultry Science. 2014; 93:1883-1889. DOI: 10.3382/ps.2013-03470

[14] Fanatico AC, Mench JA, Archer GS, Liang Y, Brewer Gunsaulis VB, Owens CM, Donoghue AM. Effect of outdoor structural enrichments on the performance, use of range area, and behavior of organic meat chickens. Poultry Science. 2016; 95: 1980-1988. DOI: 10.3382/ps/pew196

[15] Tambini A, Alba M, Perales R, Falcón N. 2010. Evaluación anatomo histopatológica de bursa, timo y bazo de pollos de carne criados sobre cama reutilizada vs. cama nueva. Rev Inv Vet Perú. 2010; 21: 180-186. DOI:10.15381/rivep. v21i2.135

[16] Jiménez-Moreno E, González-Alvarado JM, González-Sánchez D, Lázaro R, Mateos GG. Effects of type and particle size of dietary fiber on growth performance and digestive traits of broilers from 1 to 21 days of age. Poultry Science. 2010;89: 2197-2212. doi:10.3382/ps.2010-00771

[17] Dal Bosco A, Mugnai C, Mattioli S, Rosati A, Ruggeri S, Ranucci D, Castellini C. Transfer of bioactive compounds from pasture to meat in organic free-range chickens. Poultry Science. 2016; 95: 2464-2471. doi: 10.3382/ps/pev383

[18] Ponte PI, Mendes I, Quaresma M, Aguiar MN, Lemos JP, Ferreira LM, Soares MA. Cholesterol levels and sensory characteristics of meat from broilers consuming moderate to high levels of alfalfa. Poult Science 2004; 83: 810-814. DOI:10.1093/ps/83.5.810

[19] Dong XF, Gao WW, Tong JM, Jia HQ, Sa RN, Zhang Q. Effect of polysavone (alfalfa extract) on abdominal fat deposition and immunity in broiler chickens. Poultry Science 2007; 86: 1955-1959. DOI:10.1093/ps/86.9.1955

[20] Fletcher DL. A method for estimating the relative degree of saponification of xanthophyll sources and feedstuffs. Poultry Science. 2006. 85: 866-869. DOI:10.1093/ps/85.5.866

[21] Kwak H, Austic RE, Dietert RR. Influence of dietary arginine concentration on lymphoid organ growth in chickens. Poultry Science. 1999; 78:1536-1541. DOI:10.1093/ps/78.11.1536

[22] Sato K, Takahashi K, Tohno M, Miura Y, Kamada T, Ikegami S, Kitazawa H. Immunomodulation in gut-associated lymphoid tissue of neonatal chicks by immunobiotic diets. Poultry Science. 2009; 88: 2532-2538. DOI: 10.3382/ps.2009-0029

[23] Paredes M, Vásquez B. Growth, carcass characteristics, weight of internal organs and meat proximate composition of six genotypes in chickens reared in Andean region of northern Peruvian. 2020. Scientia Agropecuaria 11(3): 365-374. DOI: 10.17268/sci.agropecu.2020.03.08

[24] Laudadio V, Nahashon SN, Tufarelli V. Growth performance and carcass characteristics of guinea fowl broilers fed micronized-dehulled pea (Pisum sativum L) as a substitute for soybean meal. Poultry Science. 2012; 91: 2988-2996. DOI:10.3382/ps.2012-02473

[25] González-Alvarado JM, Jiménez-Moreno E, Lázaro R, Mateos GG. Effect of type of cereal, heat processing of the cereal, and inclusion of fiber in the diet

on productive performance and digestive traits of broilers. Poultry Science. 2007; 86: 1705-1715. DOI:10.1093/ps/86.8.1705

[26] Guzmán P, Saldaña B, Kimiaeitalab MV, García J, Mateos GG. Inclusion of fiber in diets for brown-egg laying pullets: effects on growth performance and digestive tract traits from hatching to 17 weeks of age. Poultry Science. 2015; 94: 2722-2733. DOI:10.3382/ps/pev288

[27] Kimiaeitalab MV, Cámara L, Mirzaie-Goudarzi S, Jiménez-Moreno E, Mateos GG. Effects of the inclusion of sunûower hulls in the diet on growth performance and digestive tract traits of broilers and pullets fed a broiler diet from zero to 21 d of age. A comparative study. Poultry Science. 2017; 96: 581-592. DOI:10.3382/ps/pew263

[28] Xu Y, Stark CR, Ferket PR, Williams CM, Auttawong S, Brake J. Effects of dietary coarsely ground corn and litter type on broiler live performance, litter characteristics, gastrointestinal tract development, apparent ileal digestibility of energy and nitrogen, and intestinal morphology. Poultry Science. 2015; 94: 353-361. DOI:10.3382/ps/peu016

[29] MINAGRI. 2020. Boletín Estadístico Mensual de la Producción y Comercialización Avícola [Internet]. 23 p. Available from: http://siea.minagri.gob.pe/siea/sites/ default/files/ produccion-comercializacion-avicolaene2020-100320.pdf

[30] OCDE/FAO. OCDE-FAO Perspectivas Agrícolas 2017-2026. Ed. OCDE, París. 149 pp

[31] Fanatico AC, Pillai PB, Emmert JL. Meat Quality of Slow- and Fast-Growing Chicken Genotypes Fed Low- Nutrient or Standard Diets and Raised Indoors or with Outdoor Access. Poultry Science 2007; 86: 2245-2255

[32] Funaro A, Cardenia M, Petracci S. Comparison of meat quality characteristics and oxidative stability between conventional and freerange chickens. Poultry Science. 2014; 93: 1511-1522

[33] Jaturasitha S, Srikanchai T, Kreuzer M. Differences in Carcass and Meat Characteristics between Chicken Indigenous to Northern Thailand (Black-Boned and Thai Native) and Imported Extensive Breeds (Bresse and Rhode Island Red). Poultry Science. 2008; 87: 160-169

[34] Haunshi S, Niranjan M, Shanmugam M. Characterization of two Indian native chicken breeds for production, egg and semen quality, and welfare traits. Poultry Science. 2011; 90: 314-320

[35] Osman S, Yonezawa T, Nishibori M. Origin and genetic diversity of Egyptian native chickens based on complete sequence of mitochondrial DNA D-loop región. Poultry Science. 2016; 95: 1248-1256

[36] Laenoi W, Kunkalw W, Buranawit K. Phenotypic Characterization and Farm Management of Indigenous Chicken Reared in Highland Region of Northern Thailand. Agriculture and Agricultural Science Procedia. 2015; 5: 127-132

[37] Fathi MM, Al-Homidan I, Motawei MI. Evaluation of genetic diversity of Saudi native chicken populations using microsatellite markers. Poultry Science, 2017; 96: 530-536.

[38] Riztyan R, Katano T, Shimogiri T. Genetic diversity and population structure of Indonesian native chickens based on single nucleotide polymorphism markers. Poultry Science. 2011; 90: 2471-2478.

[39] Bungsrisawat P, Tumwasorn S, Loongyai W. Genetic parameters of some carcass and meat quality traits in Betong chicken (KU line). Agriculture and Natural Resources. 2018; 52: 274-279

[40] Promket D, Ruangwittayanusorn K, Somchan T. The Study of Carcass Yields and Meat Quality in Crossbred Native Chicken (Chee). Agriculture and Agricultural Science Procedia. 2016; 11: 84-89.

[41] Rajkumar U, Haunshi S, Paswan C. Characterization of indigenous Aseel chicken breed for morphological, growth, production, and meat composition traits from India. Poultry Science. 2017; 96: 2120-2126.

[42] Manyeula F, Mlambo V, Marume U. Partial replacement of soybean products with canola meal in indigenous chicken diets: size of internal organs, carcass characteristics and breast meat quality. Poultry Science. 2020; 99: 254-262

[43] Dalle Zotte A, Ricci R, Cullere M. Effect of chicken genotype and white striping-wooden breast condition on breast meat proximate composition and amino acid profile. Poultry Science. 2020; 99: 1797-1803.

[44] Rizzi C, Contiero B, Cassandro M. Growth patterns of Italian local chicken populations. Poultry Science. 2013; 92: 2226-2235.

[45] Zanetti E, De Marchi M, Abbadi M. Variation of genetic diversity over time in local Italian chicken breeds undergoing in situ conservation. Poultry Science. 2011; 90: 2195-2201.

[46] Franco D, Rois D, Vázquez JA. Breed effect between Mos rooster (Galician indigenous breed) and Sasso T-44 line and finishing feed effect of commercial fodder or corn. Poultry Science. 2012; 91: 487-498.

[47] Duah KK, Essuman EK, Boadu VG. Comparative study of indigenous chickens on the basis of their health and performance. Poultry Science. 2020 (In press)

[48] Soglia F, Mudalal S, Babini E. Histology, composition, and quality traits of chicken Pectoralis major muscle affected by wooden breast abnormality. Poultry Science. 2016; 95: 651-659.

[49] Nalle CL, Ravindran V, Ravindran G. Nutritional value of narrow-leafed lupin (Lupinus angustifolius) for broilers. British Poultry Science. 2011; 52: 775-781. DOI: 10.1080/00071668.2011.639343

[50] Farrell DJ, Perez-Maldanado RA, Mannion PF. Optimum inclusion of peas, faba beans, chick peas and sweet lupins in poultry diets. II. Broiler experiments. British Poultry Science. 1999; 40: 674-680. DOI: 10.1080/00071669987061

[51] van Barneveld RJ. Understanding the nutritional chemistry of lupin (Lupinus spp.) seeds to improve livestock production efficiency. Nutrition Research. 1999; 12: 203-230. DOI: 10.1079/095442299-108728938

[52] Zdunczyk Z, Juskiewicz J, Frejnagel S, Gulewicz K. Influence of alkaloids and oligosaccharides from white lupin seeds on utilization of diets by rats and absorption of nutrients in the small intestine. Animal Feed Science Technology. 1998; 72: 143-154. DOI: 10.1016/S0377-8401(97)00173-9

[53] Gutiérrez A, Infantes M, Pascual G, Zamora J. Evaluación de los factores en el desamargado de tarwi (Lupinus mutabilis Sweet). Agroindustrial Science. 2016; 6: 145-149

[54] Mikulski D, Zdunczyk Z, Juskiewicz J, Rogiewicz A, Jankowski J. The effect of different blue lupine (L. angustifolius) inclusion levels on

gastrointestinal function, growth performance and meat quality in growing-finishing turkeys. Animal Feed Science Technology. 2014. 198:347-352. DOI: 10.1016/j.anifeedsci.2014.10.005

[55] Zdunczyk Z, Jankowski J, Mikulski D, Mikulska M, Lamparski G, Slominski BA. Juskiewicz J. Growth performance, gastrointestinal function and meat quality in growing-finishing turkeys fed diets with different levels of yellow lupine (L. luteus) seeds. Archive Animal Nutrition. 2014; 68: 211-226

[56] Samulikowska S, Konieczka P, Czer-winski J, Mieczkowska A, Jankowiak J. 2014. Feeding broiler chickens with particle diets containing lupin seeds (L. angustifolius or L. luteus): effect of incorporation level and mannanase supplementation on growth performance, digesta viscosity, microbial fermentation and gut morphology. Animal Feed Science. 2014; 23: 64-72

[57] Rodríguez A. Evaluación «in vitro» de la actividad antimicrobiana de los alcaloides del agua de cocción del proceso de desamargado del chocho (*Lupinus mutabilis* Sweet). [Tesis]. Riobamba, Ecuador: Escuela Superior Politécnica de Chimborazo. 2009

[58] Zeng QF, Bai P, Wang JP, Ding XM, Luo YH, Bai SP, Xuan Y, Su ZW, Lin SQ, Zhao LJ, Zhang KY. The response of meat ducks from 15 to 35 d of age to gossypol from cottonseed meal. Poultry Science. 2015; 94: 1277-1286. DOI: 10.3382/ps/pev070

[59] Paredes M, De la Flor E, Mantilla J. Effects of four dietary levels of lupine seed meal (Lupinus mutabilis) on productive parameters, intestinal development and hematological values in eight-week-old turkeys. Rev Inv Vet Perú 2019; 30(4): 1527-1536. http://dx.doi.org/10.15381/rivep.v30i4.17172

[60] INEI (Instituto Nacional de Estadística e Informática). Perú: Crecimiento y distribución de la población, 2017. Primeros Resultados. Available from: https://www.inei.gob.pe/media/MenuRecursivo/publicaciones_digitales/Est/Lib1530/libro.pdf

[61] Scanes CG, Hurst K, Thaxton Y. Effect of transportation and shackling on plasma concentrations of corticosterone and heterophil to lymphocyte ratios in market weight male turkeys in a commercial operation. Poultry Science. 2020; 99: 546-554

[62] Kalmar ID, Vanrompay D, Janssens GPJ. Broiler ascites syndrome: Collateral damage from efficient feed to meat conversion. The Veterinary Journal. 2013; 197: 169-174.

[63] Izadinia M, Nobakht M, Khajali F. Pulmonary hypertension and ascites as affected by dietary protein source in broiler chickens reared in cool temperature at high altitudes. Animal Feed Science and Technology. 2010; 155: 194-200.

[64] Boerboom G, van Kempen T, Navarro-Villa A. Unraveling the cause of white striping in broilers using metabolomics. Poultry Science. 2018; 97: 3977-3986.

[65] Druyan S, Levi E, Shinder D. Reduced O2 concentration during CAM development - Its effect on physiological parameters of broiler embryos. Poultry Science. 2012; 91: 987-997

[66] Balog JM, Anthony NB, Cooper MA. Ascites Syndrome and Related Pathologies in Feed Restricted Broilers Raised in a Hypobaric Chamber. Poultry Science. 2000; 79: 318-323

[67] Macedo-Márquez A. La producción de especies reactivas de oxígeno (EROs) en las mitocondrias de Saccharomyces cerevisiae. Revista Especializada en

Ciencias Químico-Biológicas. 2012;15(2): 97-103

[68] Young JF, Stagsted J, Jensen SK. Ascorbic Acid, α-Tocopherol, and Oregano Supplements Reduce Stress-Induced Deterioration of Chicken Meat Quality. Poultry Science. 2003; 82: 1343-1351

[69] Leskovec J, Levart A, Peric L. Antioxidative effects of supplementing linseed oil enriched diets with α-tocopherol, ascorbic acid, selenium, or their combination on carcass and meat quality in broilers. Poultry Science 2019; 98: 6733-6741.

[70] Saiz del Barrio A, Mansilla WD, Navarro-Villa A. Effect of mineral and vitamin C mix on growth performance and blood corticosterone concentrations in heat-stressed broilers. J. Appl. Poult. Res. 2020; 29: 23-33

[71] Huff GR, Huff WE, Balog JM. Stress Response Differences and Disease Susceptibility Reflected by Heterophil to Lymphocyte Ratio in Turkeys Selected for Increased Body Weight. Poultry Science. 2005; 84: 709-717

[72] Zeng QF, Yang X, Zheng P. Effects of low ambient temperatures and dietary vitamin C supplementation on pulmonary vascular remodeling and hypoxic gene expression of 21-d-old broilers. Journal of Integrative Agriculture. 2016; 15(1): 183-190.

[73] Bautista-Ortega J, Cortes-Cuevas A, Ellis. Supplemental L-arginine and vitamins E and C preserve xanthine oxidase activity in the lung of broiler chickens grown under hypobaric hypoxia. Poultry Science. 2014; 93: 979-988

Section 2

Ongoing Research in Meat and Nutrition

Chapter 4

Vitamins and Minerals in Raw and Cooked Donkey Meat

Paolo Polidori, Paola Di Girolami and Silvia Vincenzetti

Abstract

Human health is deeply affected by nutrition. The most important nutritional property of a good diet, able to provide an adequate amount of nutrients, to fulfill growth and development requirements, permitting also health maintenance, is variety of foods. Meat can be included in several diets, particularly when they are based on a restricted choice of plant foods. The inclusion of meat and meat products, even in small amounts, can significantly improve many diets; in fact, meat and derived products are good sources of proteins, vitamins and mineral salts. Thermal processes used for cooking meats represent an important factor which affects the minerals and vitamins meat content. Loss of minerals and water-soluble vitamins in cooked meat may occur, depending on the cooking method used. Previous studies investigated on donkey meat nutritional properties, described interesting characteristics of this alternative red meat, rich in protein and in iron, and with low-fat content. This chapter describes the donkey meat chemical composition, showing a comparison with other traditional red meats. The effects of cooking methods on donkey meat vitamins and minerals content will be also evaluated.

Keywords: donkey, meat quality, vitamins, minerals, cooking methods

1. Introduction

Donkey's domestication started about 10,000 years ago [1]; this animal has been particularly important for humans, because it shows a great tolerance for hard work, a strong resistance to disease and a little maintenance requirement. Donkey is an animal strictly correlated to human history; it is possible to find texts mentioning donkey in literature, in religious books and in ancient poems. Because he is stoic, slow, and sometimes stubborn, donkey has not been considered a smart animal. Throughout human history, donkey has been sometimes revered, sometimes reviled or ridiculed. Donkey has been a constant presence since the earliest human societies formed. Their greatest use to humans has been as beasts of burden, from Ancient Egypt to present day in many countries both in Africa and/or in Asia, transporting several kinds of goods on their backs or pulling small carts.

The process of donkey's domestication is still not completely known; surely donkeys were first domesticated by ancient cattle herders because of the climatic changes, with lands becoming more arid and rainfall more unpredictable. Probably they have been domesticated several times by different groups of herders, while interbreeding among wild and domestic asses is continued throughout the entire domestication process [2]. Donkeys were bred mainly as working animals; livestock production, such as both meat and milk, were basically considered by-products.

Donkey domestication represents a crucial step in human history, creating a new phase in human populations, moving from a sedentary lifestyle to a new society with an economy based on trade.

2. Donkey breeding today

Actually, the total amount of donkeys all-over the world is about 44 million (**Table 1**), mostly in developing countries [3]. There are over 189 donkey breeds, showing body weights ranging from 80 to 480 kg, and heights at withers ranging between 64.2 cm of the smallest breed (Mediterranean Miniature Donkey) and 170 cm of the tallest breed (American Mammoth Jackstock).

In Europe 60 donkey breeds are actually classified, the most represented are listed in **Table 2**; however, this total amount cannot be considered fully accurate. In fact, among the 60 breeds listed, two of them must be unfortunately considered extinct, while five other breeds are basically synonymous. For six other breeds, only old data about their consistency are shown, while for other five breeds, the actual consistency is reported, but the description of phenotype or photos are not provided. For 14 breeds any values have been reported. Basically, only for 28 breeds morphologic description is really shown and the updated consistency is also reported [5].

2.1 Donkey meat production

Donkeys are normally not considered as meat producing animals. Bovines, buffaloes, and camels are usually bred as working animals, as well as for milk and meat production. On the other hand, among equids horse meat is considered a good alternative red meat, with specific dietary characteristics; its consumption registered in some countries, such as Spain, France, Poland, a significant increase in recent years [6]. Horse meat chemical composition is in fact characterized by low total fat and cholesterol content, a healthy ratio between saturated and unsaturated fatty acids and a good minerals content, especially iron and zinc [7]. Consumers, being more health conscious, actually look for foods with high standard quality, requiring leaner meat, with basically the minimal fat content necessary to maintain adequate sensorial properties, taking also into consideration meat tenderness, juiciness and flavor. The success of any food is directly correlated with consumer's acceptance. Meat quality as perceived by the consumers is strictly associated to its physical and chemical composition, even if quality requirements for meat acceptability by the consumers are not the same all-over the world.

Continent	Donkeys
Asia	15,000,000
Africa	9,700,000
Middle East	9,220,000
South America, Caribbean	8,164,000
Europe	1,500,000
North America	52,000
TOTAL	43,636.000

Table 1.
World donkeys distribution [4].

Country	Breed	Consistency
Croatia	Littoral dinaric	2,150
France	Ane de Provence	271
	Ane Normand	221
Italy	Ragusano	2,481
	Martina Franca	1,086
Portugal	Buro de Miranda	1,400
Spain	Catalana	957
	Zamorano-Leonés	1,338

Table 2.
Main donkey breeds in Europe [5].

Today China is the top producer of donkey meat in the world, followed by Niger (**Table 3**). Considering the well-known good reputation associated to horse meat, the aim of this chapter is to show the recent findings about donkey meat quality, describing its nutritional properties compared to other red meats, evaluating the effects of cooking methods on losses of minerals and vitamins compared to the original amount of these micronutrients in raw donkey meat.

2.2 Donkey meat quality

Donkeys can be considered an important meat producing animal in arid and semi-arid areas, but donkey breeding for meat production does not represent an important activity for farmers, because donkey meat has a bad reputation. In fact, in the countries in which donkeys are still used as working animals in agriculture, they are normally slaughtered at an advanced age, producing a meat unacceptably tough [9]. However, not all male donkeys can be used as stallion for reproduction, while breeding young males for meat production is an easy way to produce a cheap food, helping local farmers in increasing their own income. In fact, donkey is an animal with a great ability in surviving in extreme weather and environmental conditions, such as high temperatures, low rainfall and poor feed availability. So, donkey can be bred as a meat producing animal in regions where the weather conditions and the feed availability negatively affect other animal's production efficiency.

The term 'meat' is normally associated only to meat flesh, basically skeletal muscle plus the possible attached connective tissue or fat; the term offal describes

Country	Number of donkeys	Meat production (tons/year)
China	2,249,807	183,755
Niger	124,319	9,946
Burkina Faso	72,372	4,342
Senegal	52,581	3,155
Mali	50,598	3,036
Mauritania	22,920	2,521
WORLD	2,569,520	207,172

Table 3.
Donkey meat production in the world [8].

	Beef [11]	Lamb [11]	Pork [11]	Horse [11]	Donkey [12]
Water	71.6	70.2	66.0	75.2	73.7
Protein	20.3	19.8	20.8	21.1	22.8
Fat	8.1	9.9	13.2	3.7	3.5

Table 4.
Chemical composition (g/100 g) of muscle Longissimus thoracis *taken from domestic mammals.*

meat products different than meat flesh, such as brain, heart, kidney, liver, pancreas, spleen, thymus, tongue and tripe [10]. The term 'red meat' is traditionally associated to meat from beef, lamb and goat, even if equid meat can also be considered another kind of red meat [8]. In **Table 4** is shown the chemical composition determined in meat samples taken from beef, lamb, pork, horse and donkey; protein content is very similar in all the meats, while fat content is significantly lower in horse and donkey meat.

3. Meat as a source of vitamins

Meat shows an interesting content of all the B-complex vitamins, such as thiamine (vitamin B_1), riboflavin (vitamin B_2), niacin (vitamin B_3), biotin (vitamin B_8), pyridoxine (vitamin B_6), cobalamin (vitamin B_{12}), pantothenic acid (vitamin B_5) and folic acid (vitamin B_9). The last two show higher contents particularly in liver, which is rich in vitamin A and provides interesting amounts of the other fat-soluble vitamins D, E and K.

Red meat provides about 25% of the recommended dietary intakes for riboflavin, niacin, vitamin B_6, pantothenic acid and almost 70% of the daily requirement of vitamin B_{12} in 100 g of serving [10]. Lack of vitamin B_{12}, together with iron deficiency, is the main cause of anemia in human population.

Vitamins are chemically reactive molecules. Cooking methods affects the stability of several vitamins, causing significant losses in cooked meat. Other foods show significant decrease in vitamins content caused by thermal treatments. In vegetables, vitamin C and thiamine are the most susceptible to cooking degradation, while in fresh milk pasteurization caused a 25% loss of Vitamin C, which increased up to 60% after UHT treatment [13].

Vitamins B-complex is well represented in equid meat, too (**Table 5**). Horse and donkey meat have been recently investigated, showing nutritional properties remarkably similar to the most common red meats, such as beef, lamb and buffalo [14].

Vitamin	Horse meat	Donkey meat
Thiamine (mg/100 g)	0.04	0.02
Riboflavin (mg/100 g)	0.18	0.11
Niacin (mg/100 g)	5.54	4.03
Vitamin B_6 (mg/100 g)	0.64	0.48
Vitamin B_{12} (μg/100 g)	2.08	1.90

Table 5.
B-vitamins in equid meat [8].

4. Meat as a source of minerals

Living organisms need minerals to maintain a healthy condition. In fact, lack or deficiency of trace elements can produce diseases that are reversed once incorporated [15]. Several of the essential trace elements are metals: copper (Cu), zinc (Zn), iron (Fe), and manganese (Mn), all of them involved in different metabolic reactions. Iron is also involved in the metabolism of the oxygen carrier hemoproteins, namely hemoglobin and myoglobin [16]. Humans obtain the recommended micronutrients intakes through diet [17].

Meat is the richest source of zinc in the diet and supplies one third to one half of the total zinc recommended daily intake. A zinc dietary deficiency has been detected in teenagers in the Middle East eating a poor diet mainly based on unleavened bread. Red meats such as beef and lamb are the richest sources of the minerals iron and zinc; a serving size of 100 g of red meat provides about 25% of daily adult requirements [10]. The iron in meat is mostly heme iron, which is well absorbed. Also zinc absorption is higher in animal source foods compared to plant foods; zinc requirements may be 50% higher for vegetarians. Red meats are also good sources of selenium, providing more than 20% of daily intake requirement. Lean meat is low in sodium, with a ratio potassium/sodium >5. The copper content in raw lean meat is in the range 0.055–0.190 mg/100 g in beef and veal, 0.090–0.140 mg/100 g in lamb, and 0.190–0.240 mg/100 g in mutton [10].

Minerals content has also been determined in both horse and donkey meat (**Table 6**); values obtained confirm the interesting nutritional properties of equid meat [19] compared to other traditional red meats.

5. Cooking meat

Cooking meat can guarantee food safety to the consumer. Meat can be cooked at high temperatures for long time, but this cooking method can affect meat nutritional and sensorial quality, with loss of water-soluble vitamins and thermolabile minerals [20].

Minerals	Horse [8]	Donkey [18]	Beef [10]	Lamb [10]	Veal [10]
Macroelements					
Potassium	191–203	312–438	363	344	362
Phosphorus	186–198	185–335	215	194	260
Sodium	52.6–68.1	36.8–83.6	51	69	51
Magnesium	18.5–33.6	38.7–43.3	25	28	26
Calcium	4.11–4.51	6.12–11.5	4.5	7.2	6.5
Microelements					
Iron	2.56–4.04	2.86–4.77	1.8	2.0	1.1
Zinc	2.07–2.66	2.99–4.71	4.6	4.5	4.2
Copper	0.13–0.21	N.D.	0.12	0.12	0.08
Manganese	0.01–0.002	N.D.	N.D.	N.D.	N.D.

Table 6.
Mineral content (mg/100 g) in different meats.

Meat proteins are differently affected by cooking temperatures. In the case of boiling or microwave cooking, temperature gets over 100 °C, causing proteins denaturation due to an enzymatic inactivation of lipases, proteases, etc. When cooking temperature ranges between 100 and 140 °C, that is the situation related to pressure cooking and baking, meat digestibility is reduced because of the formation of intramolecular and intermolecular covalent bonds [21]. In fried and roasted meat, when cooking temperatures are above 140 °C, amino acid destruction occurs, particularly cysteine or tryptophan, followed by an isomerization to D-configuration and a consequent reduction of the nutritional value [22].

Meat fat can undergo toward a fusion process after heat treatment, but it is hard to determine the exact melting point, because of the different triglyceride mixtures; depending on the type of fat, in fact, triglycerides decompose at a different temperature [23]. High cooking temperatures can sometimes form toxic cyclic monomers, dimers and polymers; this is the typical situation that occurs when acrolein is formed after high thermal treatments [24].

Purchased meat may include bone, external fat, gristle and tendons which are often removed before cooking, so that the meat composition "on the plate" can be quite different. Meat and meat products are considered cooked when the internal temperature is maintained at 65–70 °C for 10 minutes, with a consequent coagulation of the proteins and a meat tenderizing effect due to the partial hydrolysis of the collagen [25]. Most of meat bacteria will be destroyed during cooking, only thermoresistant spores can survive above 100 °C [26]. The end of the cooking process can be normally detected by a change of color from red to brown with an associated flavors development [27]. In fact, cooked flavor can be considered the result of several reactions affecting changes in fat and proteins, with heat breakdown of peptides and amino acids [28].

Meat obtained from animals slaughtered at older ages, as normally occurs in donkey breeding, is richer in connective tissue, therefore longer cooking times at 50–60 °C are necessary, because at this temperature collagen can be hydrolyzed [29]. If heated for long times at temperature above 80 °C, unpleasant flavors can be produced because amino acids begin to decompose; hydrolysis of collagen, in fact, is fast when high temperatures are used for only a short time [30]. Finally, water is lost during cooking, too. The loss of water is strictly correlated with cooking time, cooking temperature, cooling method, serving size, heat penetration and original meat chemical composition. Loss of water causes an increase in fat and protein concentration in cooked meat [27].

6. Effects of cooking methods on meat vitamins content

Meat cooking techniques show significant effects on vitamins content, especially considering the important losses of B vitamins complex [31]. Vitamin B_{12} and thiamine are the most affected B vitamins, while riboflavin and niacin show lower decreases [32]. B vitamins are water soluble and thermally instable, therefore some cooking methods, such as boiling, may provoke high vitamin losses, while shorter cooking time may reduce these losses [33].

In the juices exuded from meat during cooking are contained small amount of all the water-soluble constituents, minerals, proteins and vitamins. Losses of thiamine are normally in the range 15–40% in boiled meat, 40–50% in fried meat, 30–60% in roasted meat and 50–70% in canned meat. Average cooking losses of riboflavin are close to 10%. Riboflavin is relatively stable to most cooking methods, excluding the high temperature necessary for roasted meat. Canning and dehydration do not significantly affect riboflavin content, either. Niacin is quite stable to several thermal treatments: losses normally average about 10% in cooked meat [33].

There are not many data available about other B vitamins: according to a recent report, about 30% of the vitamin B_6 and pantothenate are lost in cooked meat [34].

7. Effects of cooking methods on meat minerals content

Minerals loss during meat cooking occurs, too; the amounts of these nutrients ingested with cooked meat could show great variations compared with their content in raw meat. For example, heme iron is converted after different thermal treatments in non-heme iron, the less available form of this mineral [35].

The most common methods normally used in order to determine trace elements content in foods are based on microwave-assisted digestion with mineral acids followed by a measurement using atomic spectrometry techniques [15].

Cooking process reduces the meat content of the trace elements like zinc, magnesium, iron, phosphorous; some of these nutrients can be completely lost after thermal treatments [36]. A study performed by Kimura & Itokawa [37] determined in home cooked hamburger an average loss of 56% of total minerals content, with the following values for each macro and micronutrients:

- Sodium (Na): - 54%
- Potassium (K): - 67%
- Phosphorus (P): - 66%
- Calcium (Ca): - 47%
- Magnesium (Mg): - 59%
- Iron (Fe): - 46%
- Zinc (Zn): - 71%
- Copper (Cu): - 41%

In the same study [37] cooking loss of minerals in pork was largest in zinc, followed by sodium, potassium, calcium, iron, magnesium and phosphorus. In these samples, cooking loss was correlated with the different cooking method, not on the type of mineral. A recent study [15] was performed with the aim of determining minerals content in beef cooked using two levels of doneness, respectively medium and well done. The results obtained for medium cooking level showed this range of minerals content (μg/100 g serving size):

- Cu (84.9–117.4);
- Fe (2288–2689);
- Mn (11.6–20.7);
- Ni (12.4–19.8);
- Zn (4100–7471);

For well-done cooking level the results obtained were in the following ranges:

- Cu (76.3–97.0);
- Fe (1886–2689);
- Mn (9.7–15.4);
- Ni (12.7–26.8);
- Zn (3187–6204).

No significant minerals loss was detected in meat when medium cooking level was applied and total amount of exuded juices caused a weight loss of 16%, but when the weight loss was 28–35%, following the well-done cooking level, minerals content significantly decreased in cooked meat.

8. Effects of cooking methods on donkey meat

A study has been performed at the University of Camerino, Italy, in order to detect minerals and vitamins B complex content in raw meat obtained slaughtering donkey males 20 months old. The influence of donkey meat cooking process on the retention of these micronutrients was also evaluated.

Twelve male entire crossbred (Romagnolo x Amiata) donkeys born and reared in the same farm in extensive conditions based on pasture were slaughtered at 20 months of age, with an average final body weight of 246 ± 20 kg. After slaughtering, all the carcasses were transferred to a cold room at a temperature of 4 °C; after 24 h, four samples of 600 g were taken from the muscle Longissimus thoracis (LT) at the height of the 12-14th thoracic vertebra.

From each carcass, two samples (300 g) of muscle LT were used for raw meat chemical analysis, the other two LT samples (300 g) were cooked in an oven at 170 °C for 45 min, following a common type of home cooking. Both raw and cooked meats were freeze-dried and then ground in a food blender to ensure homogeneity and representative samples for analysis [38]. B complex vitamins were quantified by HPLC after acidic and enzymatic hydrolysis of the samples [39].

To assess the minerals content in raw and cooked donkey meat was used a published guide [40] for calculating the mineral retention percentages in foods prepared using different cooking methods, normalizing a serving size portion of 100 g to the weight loss during cooking [31]. Macro (Ca, K, Mg, Na, P) and microelements (Cu, Mn, Fe, Zn) were mineralized using the Mileston Ethos 900 microwave; samples were analyzed by means of atomic absorption spectroscopy (AAS) equipped with a low-flow Gem Cone ultrasonic nebulizer for the detection of very low concentrations [41].

8.1 Vitamins and minerals losses in cooked donkey meat

Vitamins content in raw and cooked donkey meat is shown in **Table 7**. Niacin content was the most abundant vitamin determined in raw meat, followed by pantothenic acid, vitamin B_5, then pyridoxine, vitamin B_6. Thiamine content was higher compared to the same vitamin level determined in beef (0.04 mg/100 g), but lower compared to the content (0.12 mg/100 g) detected in lamb [10]. Higher contents of niacin (7.3 mg/100 g) and thiamine (0.16 mg/100 g) were found in horse fillet,

	Raw Donkey Meat	Cooked Donkey Meat
Thiamine – Vitamin B_1 (mg/100 g)	0.09 ± 0.01	Trace
Riboflavin – Vitamin B_2 (mg/100 g)	0.22 ± 0.07	0.18 ± 0.05
Niacin -Vitamin B_3 (mg/100 g)	6.09 ± 0.27	5,22 ± 0.16
Pantothenic acid – Vitamin B_5 (mg/100 g)	1.13 ± 0.23	0.99 ± 0.06
Piridoxine – Vitamin B_6 (mg/100 g)	0.61 ± 0.12	0.49 ± 0.08
Cobalamine – Vitamin B_{12} (μg/100 g)	1.80 ± 0.15	1.10 ± 0.04

Table 7.
Vitamin B complex content in raw and cooked donkey meat.

while riboflavin content was lower (0.20 mg/100 g) in this kind of meat, confirming the great variability in meat vitamins content among species [31]. Vitamin B_{12} level in raw donkey meat is close to the contents of this vitamin determined in different beef cuts [42].

Cooking procedure decreased B vitamins complex content, mainly thiamine, that resulted significantly reduced by thermal degradation, becoming hard to detect in the cooked meat analyzed. Niacin content showed a significant decrease after cooking, too: its retention was about 35%. Riboflavin resulted more stable to heat, showing a retention value close to 70%; these results confirmed the findings of Lombardi-Boccia et al. [31], who found in roast beef a complete riboflavin retention, a severe loss in niacin, while thiamine was basically undetectable. Vitamin B_{12} showed a significant decrease in cooked donkey meat, confirming the results obtained in cooked beef [43] and in grilled fish [44]; the Authors in both the cited studies attributed the loss of vitamin B_{12} to the destruction of this vitamin caused by the heating.

The effect of cooking procedure on minerals content in donkey meat is pointed out in **Table 8**. Cooked donkey meat did not show significant decrease in the iron content compared to raw meat; in a study performed on horse fillet [38], a significant loss in total iron content after cooking the horse meat 180 °C for 50 min was reported, confirming the effect of cooking time and temperature on minerals content. Previous studies [38, 45] demonstrated that in cooked meat total iron

	Raw Donkey Meat	Cooked Donkey Meat
Macroelements		
Calcium (Ca)	3.46 ± 0.64	3.15 ± 0.53
Potassium (K)	285.60 ± 26.8	228.17 ± 13.8
Magnesium (Mg)	24.36 ± 1.12	22.12 ± 1.23
Sodium (Na)	67.25 ± 3.31	72.84 ± 3.42
Phosphorus (P)	213.35 ± 11.4	204.43 ± 10.6
Microelements		
Copper (Cu)	0.21 ± 0.03	0,19 ± 0.05
Manganese (Mn)	0.03 ± 0.005	0.01 ± 0.004
Iron (Fe)	2.81 ± 0.38	2.39 ± 0.41
Zinc (Zn)	4.35 ± 0.13	3.90 ± 0.12

Table 8.
Mineral content (mg/100 g) determined in raw and cooked donkey meat.

content did not decrease compared to raw meat, but the heme/non-heme iron ratio was modified; the heme iron content decreased according to the different heat treatment utilized. Zinc content in donkey meat was not significantly affected by the cooking method (**Table 8**); this insoluble mineral is linked to proteins and tend to remain in the meat during cooking [27]. Among the other minerals, copper remained quite stable after cooking process, and the other minerals did not show significant differences, confirming the results obtained in a previous study [46].

9. Conclusions

Vitamins and minerals have been recognized to be particularly important in human nutrition. Meat provides significant amounts of nutrients, including thiamine, vitamin B_{12}, available iron and zinc. Physical and chemical changes in meat can be caused by different cooking methods. Meat tissue structure is affected by physical changes, regarding in particular meat sensorial characteristics such as color, texture, aroma, and taste.

The remarkable nutritional value of donkey meat is summarized in its low fat content, an interesting protein concentration together with a valid heme iron content showing that their consumption may be favorable for human health. The nutritional characteristics of donkey meat show interesting aspects in comparison to the usual red meat; when related to the human health parameters, this kind of meat can be favourably accepted by the consumers.

Production of donkey meat shows a very important potential to be considered in the economic development of many countries; there is a big challenge for the farmers to use donkeys as meat animals. Donkey, especially young males, can be used as cheap meat animals, and donkey meat can easily have a market in the western countries, considering the high quality level shown by this kind of red meat.

The loss of vitamins and minerals in cooked meat are due to the molecular interactions that occur when thermal treatment is applied. Cooking methods strongly affect trace elements and B vitamins content in meat, causing losses during cooking processes. The amount of these nutrients ingested with cooked meat could be hugely different compared to raw food, according to the different cooking methods. Cooked donkey meat showed lower values of thiamine and niacin compared to raw muscle, while riboflavin content was not significantly affected by cooking. Minerals retention in cooked donkey meat showed that zinc, copper and iron contents did not significantly decrease after thermal treatments. Donkey meat, rediscovering the local culinary tradition, can represent a valuable niche quality food in human diet.

Donkey meat production could help in taking care of many marginal lands of the world, where soil stability and animal biodiversity preservation represent hard worries. Donkeys possess a peculiar digestive physiology, being hindgut fermenter herbivores: for this reason, they can advantageously compete with big and/or small ruminants for pastures utilization. In order to fulfill this target, use of indigenous donkey breeds can help in preserving local animal biodiversity. Further studies must be performed to investigate on the effects of feeding on donkey meat quality. Moreover, no data are actually available about the presence in donkey meat of specific nutraceutical molecules, such as carnosine, carnitine and taurine.

Conflict of interest

"The authors declare no conflict of interest."

Author details

Paolo Polidori[1*], Paola Di Girolami[1] and Silvia Vincenzetti[2]

1 School of Pharmacy, University of Camerino, Camerino, MC, Italy

2 School of Biosciences and Veterinary Medicine, University of Camerino, Matelica, MC, Italy

*Address all correspondence to: paolo.polidori@unicam.it

References

[1] Bough J. Human and donkey relationships since domestication. In: Navas González FJ, editor. Current Donkey Production & Functionality: Relationship with humans. Cordoba: UCOPress; 2017. p. 40-72.

[2] Kimura B, Marshall F, Beja-Pereira A, Mulligan C. Domestic donkeys. African Archaelogical Review. 2013;30:83-95.

[3] Polidori P, Ariani A, Micozzi D, Vincenzetti S. The effects of low voltage electrical stimulation on donkey meat. Meat Science. 2016;19:160-164.

[4] Polidori P, Vincenzetti S. The Therapeutical, Nutritional and Cosmetic Properties of Donkey Milk. Cambridge: Cambridge Scholars Publishing; 2019. 154 p.

[5] Camillo F, Rota A, Biagini L, Tesi M, Fanelli D, Panzani D. The Current Situation and Trend of Donkey Industry in Europe. Journal of Equine Veterinary Science. 2018;65:44-49.

[6] Lorenzo JM, Sarriés MV, Tateo A, Polidori P, Franco D, Lanza M. Carcass characteristics, meat quality and nutritional value of horsemeat: A review. Meat Science. 2014;96:1478-1488.

[7] Belaunzaran X, Bessa RJB, Lavín P, Mantecón AR, Kramer JKG, Aldai N. Horse-meat for human consumption—Current research and future opportunities. Meat Science. 2015;108:74-81.

[8] Lorenzo JM, Maggiolino A, Sarriés MV, Polidori P, Franco D, Lanza M, De Palo P. Horsemeat: Increasing Quality and Nutritional Value. In: Lorenzo JM, Munekata PES, Barba FJ, Toldrà F, editors. More than Beef, Pork and Chicken – The Production, Processing, and Quality Traits of other Sources of Meat for Human Diet. Switzerland: Springer Nature; 2019. p. 31-67.

[9] Marino R, Albenzio M, della Malva A, Muscio A, Sevi A. Nutritional properties and consumer evaluation of donkey bresaola and salami: comparison with conventional products. Meat Science. 2015;101:19-24.

[10] Williams P. Nutritional composition of red meat. Nutrition & Dietetics. 2007;64 (Suppl. 4):S113–S119.

[11] Levine MA. Eating horses: the evolutionary significance of hippophagy. Antiquity. 1998;72:90-100.

[12] Polidori P, Vincenzetti S, Cavallucci, C, Beghelli D. Quality of donkey meat and carcass characteristics. Meat Science. 2008;80:1222-1224.

[13] Dhakal SP, He J. Microencapsulation of vitamins in food applications to prevent losses in processing and storage: A review. Food Research International, 2020;137:109326.

[14] Belaunzaran X, Lavín P, Mantecón A.R., Kramer J.K.G., Aldai N. Effect of slaughter age and feeding system on the neutral and polar lipid composition of horse meat. Animal. 2018;12:417-425.

[15] Piston M, Suarez A, Bühl V, Tissot F, Silva J, Panizzolo L (2020) Influence of cooking processes on Cu, Fe, Mn, Ni, and Zn levels in beef cuts. Journal of Food Composition and Analysis. 2020;94:103624.

[16] Jiang Z, You Q, Zhang X. Medicinal chemistry of metal chelating fragments in metalloenzyme active sites: a perspective. European Journal of Medical Chemistry. 2019;165:172-197.

[17] Whittaker JW. Metal uptake by manganese superoxide dismutase.

Biochimica et Biophysica Acta (BBA) - Proteins and Proteomics. 2010;1804:298-307.

[18] Polidori P, Cavallucci C, Beghelli D, Vincenzetti S. Physical and chemical characteristics of donkey meat from Martina Franca breed. Meat Science. 2009;82: 469-471.

[19] Sarriés MV, Beriain MJ. Carcass characteristics and meat quality of male and female foals. Meat Science. 2005;70:141-152.

[20] Gómez I, Janardhanan R, Ibañez FC, Beriain MJ. The Effects of Processing and Preservation Technologies on Meat Quality: Sensory and Nutritional Aspects. Foods. 2020;9:1416.

[21] Pingen S, Sudhaus N, Becker A, Krischek C, Klein G. High pressure as an alternative processing step for ham production. Meat Science. 2016;118:22-27.

[22] Sun S, Sullivan G, Stratton J, Bower C, Cavender G. Effect of HPP treatment on the safety and quality of beef steak intended for sous vide cooking. LWT Food Science and Technology. 2017;86:185-192.

[23] Ma Q, Hamid N, Oey I, Kantono K, Farouk M. The impact of high-pressure processing on physicochemical properties and sensory characteristics of three different lamb meat cuts. Molecules. 2020;25:2665.

[24] Diéguez PM, Beriain MJ, Insausti K, Arrizubieta MJ. Thermal analysis of meat emulsion cooking process by computer simulation and experimental measurement. International Journal of Food Engineering. 2010;6:1-21.

[25] Chen X, Tume RK, Xiong Y, Xu X, Zhou G, Chen C, Nishiumi T. Structural modification of myofibrillar proteins by high-pressure processing for functionally improved, value-added, and healthy muscle gelled foods. Critical Review of Food Science and Nutrition. 2018;58:2981-3003.

[26] Dominguez-Hernandez E, Salaseviciene A, Ertbjerg P. Low-temperature long-time cooking of meat: Eating quality and underlying mechanisms. Meat Science. 2018;143:104-113.

[27] Nikmaram P, Yarmand MS, Emamjomeh Z. Effect of cooking methods on chemical composition, quality and cook loss of camel muscle (Longissimus dorsi) in comparison with veal. African Journal of Biotechnology. 2011;10:10478-10483.

[28] Sánchez del Pulgar J, Roldan M, Ruiz-Carrascal J. Volatile compounds profile of sous-vide cooked porkcheeks as affected by cooking conditions (vacuum packaging, temperature and time). Molecules. 2013;18:12538-12547.

[29] Botinestean C, Keenan DF, Kerry JP, Hamill RM. The effect of thermal treatments including sous-vide, blast freezing and their combinations on beef tenderness of M. semitendinosus steaks targeted at elderly consumers. LWT Food Science and Technology. 2016;74:154-159.

[30] Park CH, Lee B, Oh E, Kim YS, Choi YM. Combined effects of sous-vide cooking conditions on meat and sensory quality characteristics of chicken breast meat. Poultry Science. 2020;99:3286-3291.

[31] Lombardi-Boccia G, Lanzi S, Aguzzi A. Aspects of meat quality: trace elements and B vitamins in raw and cooked meats. Journal of Food Composition and Analysis. 2005;18:39-46.

[32] Riccio F, Mennella C, Fogliano V. Effect of cooking on the concentration of Vitamins B in fortified meat

products. Journal of Pharmaceutical and Biomedical Analysis 2006;41:1592-1595.

[33] Yang J, Science N. Sensory qualities and nutrient retention of beef strips prepared by different household cooking techniques. Journal of the American Dietetic Association. 1994;94:2-5.

[34] Suleman R, Wang Z, Aadil R, Hui T. Hopkins DL, Zhang D. Effect of cooking on the nutritive quality, sensory properties and safety of lamb meat: Current challenges and future prospects. Meat Science. 2020;167:108172.

[35] Carpenter CE, Clark E. Evaluation of methods used in meat iron analysis and iron content of raw and cooked meats. Journal of Agricultural and Food Chemistry. 1995;43:1824-1827.

[36] Czerwonka M, Szterk A. The effect of meat cuts and thermal processing on selected mineral concentration in beef from Holstein-Friesian bulls. Meat Science. 2015;105:75-80.

[37] Kimura M, Itokawa Y. Cooking Losses of Minerals in Foods and Its Nutritional Significance. Journal of Nutritional Science and Vitaminology. 1990;36:S25-S33.

[38] Lombardi-Boccia G, Martinez-Dominguez B, Aguzzi A. Total heme and non-heme iron in raw and cooked meats. Journal of Food Science. 2002;67:1738-1741.

[39] Barna E, Dworschàk E. Determination of thiamine (vitamin B1) and riboflavin (vitamin B2) in meat and liver by high-performance liquid chromatography. Journal of Chromatography. 1994;668:359-363.

[40] Bognar A, Piekarski J. Guidelines for recipe information and calculation of nutrient composition of prepared food dishes. Journal of Food Composition and Analysis. 2000;13:391-410.

[41] Kadim IT, Mahgoub O, Al-Marzooqi W, Al-Zadjali S, Annamalai K, Mansour MH. Effects of age on composition and quality of muscle Longissimus thoracis of the Omani Arabian camel (Camelus dromedaries). Meat Science. 2006;73:619-625.

[42] de Castro Cardoso Pereira PM, dos Reis Baltazar Vicente AF. Meat nutritional composition and nutritive role in the human diet. Meat Science. 2013;93:586-592.

[43] Watanabe F, Abe K, Fujita T, Goto M, Hiemori M, Nakano Y. Effects of microwave heating on the loss of vitamin B12 in foods. Journal of Agricultural and Food Chemistry. 1998;46:206-210.

[44] Nishioka M, Kanosue F, Yabuta Y, Watanabe F. Loss in vitamin B12 in fish (Round Herrings) meats during various cooking treatments. Journal of Nutritional Science and Vitaminology. 2011;57:432-436.

[45] Al-Khalifa AS, Dawood AA. Effects of cooking methods on thiamin and riboflavin contents of chicken meat. Food Chemistry. 1993;48:69-74.

[46] Severi S, Bedogni G, Manzieri AM, Poli M, Battistini N. Effects of cooking and storage methods on the micronutrient content of foods. European Journal of Cancer Prevention. 1997;6 (suppl 1):521-524.

Chapter 5

Nutritional Composition of Game Meat from Wild Species Harvested in Europe

Almudena Soriano and Carlos Sánchez-García

Abstract

A discussion about the nutritional composition of game meat, with specific focus on wild species harvested in Central and Mediterranean European countries has been conducted. Given the wide range of species, and the climate and vegetation differences among the harvesting areas, game meat shows heterogeneous characteristics and chemical composition, the latter being also affected by sex, age, body condition, physiological and sexual status, and hunting period. However, there are similarities which make it clearly distinguishable from livestock meat. When considering the most consumed species (red and fallow deer, wild boar, hare and wild rabbit), their meat has low fat content (<3 g/100 g for large and <4 g/100 g for small wild game species), high protein content (20–26 g/100 g) and low energy content (90–113 kcal/100 g). Wild game meat has a healthier fatty-acids profile compared to other meats, showing a higher proportion of PUFA, especially *n*-3, and consequently more favorable PUFA/SAF ratio. Wild ruminants' meat shows a favorable *n*-6/*n*-3 ratio (lower or close to 4). It has a high content of K, followed by P and micro-minerals such as Zn and Fe, together with B-group vitamins and vitamin E. Game meat from wild species harvested in Europe can diversify the market being an alternative to others red meats owing to its nutritional quality and organoleptic characteristics.

Keywords: game meat, chemical composition, fatty acid profile, cholesterol, vitamins, minerals, health, nutrition

1. Introduction

During the last decades game meat demand has increased in Europe, mainly because consumers perceive this meat as "natural and sustainable". Animals are free range, fed mainly with natural food (i.e. pasture) and their meat is free from hormones, antibiotics and other products. Meat from wild game species, meets animal welfare ethical standards, and the rearing of the animals in the wild has a lower or null impact on the ecosystems when compared to farmed species [1–3]. Therefore, consumers perceive this meat differently from traditional meat derived from domestic species, being a seasonal product that is available in fresh during the colder seasons [1], but also consumed in other seasons after being frozen or as cured meat, including specialities such as *cecina, salchichón, chorizo,* and conserves and pates. European consumers consider game meat as innovative food and are willing

IntechOpen

to pay a higher price [2], and they are likely to increase its consumption provided that it will have a higher quality and greater commercial availability [3].

However, game meat consumption is still very low in Europe, as only 2–4% of the population consumes this type of meat regularly [4]. This can be explained by its high price, seasonal availability, the lack of habit of using recipes with game meat among consumers not related with hunting, and even safety concerns [5]. As pointed by a recent study [6], the lack of knowledge on hunting may hamper game meat consumption, as hunters and their relatives show higher rates of consumption compared to people not familiarized with this activity [5]. Thus, when aiming to increase the intake of game meat and its derived products, information should be provided to sellers and consumers about the role of hunting on wildlife conservation and rural economy, but also on the nutritional quality of game meat and its possible inclusion in a balanced and healthy diet. In this way, the European Union (EU) is promoting the consumption of game meat in some countries through the program "European wild meat, nature at its purest" [7]. In France, about half of the game meat producers promote the "French game meat brand", with a strong promotion campaign targeting mainly chefs and consumers, highlighting the meat quality and its gastronomic potential [8]. In Spain, the National Association of Game Meat Producers (ASICCAZA) is conducting a programme to increase the game meat profile among consumers [9].

Meat from game species can be produced from farmed or wild animals. In some countries such as New Zealand, Australia, China and Canada, meat from farmed game is considered as an important meat subsector, and the rearing of farmed game is also growing in European countries. However, hunting of wild game remains as an important activity in Europe, including ungulates, lagomorphs and birds, being this the case of Spain, where approximately 20,000 tons are produced each year, with an estimated value of 45 million € [10].

Regulation (EC) No 853/2004 considered wild game meat as meat obtained from wild birds, ungulates and lagomorphs, as well as other land mammals that are hunted for human consumption, including animals living in enclosed territory under conditions of freedom similar to those of wild game. Among the most hunted species in Europe are wild ungulates, including red deer (*Cervus elaphus*),

Figure 1.
Main wild game species harvested in Central and Mediterranean European countries.

fallow deer (*Dama dama*), roe deer (*Capreolus capreolus*) and wild boar (*Sus scrofa*), together with birds such as ring-necked pheasants (*Phasianus colchicus*), partridges (*Alectoris spp., and Perdix spp.*), waterfowl and wader species but also mammals such as wild rabbit (*Oryctolagus cuniculus*) and hares (*Lepus spp*.) (**Figure 1**). Meat from the majority of hunted species can be legally sold in the European marked, as long it has passed a proper official *post-mortem* inspection at game-handling establishments as cutting plants, as specified with the European regulation (EC) No 853/2004.

2. Factors affecting chemical composition and characteristics of wild game meat

As game meat is produced from different groups of species (birds and mammals), and from different species within the same taxonomic category, there are considerable differences on quality attributes, including physicochemical and sensorial characteristics, together with microbiology, making game meat a heterogeneous product. In addition, quality attributes can be difficult to standardize and control since game meat from species hunted in the wild is affected by several *ante-mortem* and *post-mortem* factors.

2.1 Ante-mortem factors

Several *ante-mortem* factors affect chemical meat composition, including specie, sex, age, body condition, muscle type, hormone levels, site, climate, harvesting period and type of hunting [11, 12].

In Europe, hunting is practised all over the continent; hence a same species can be harvested in different environmental conditions, which implies different climate and habitat resulting in differences on types of food and its availability. However, for the majority of game species, hunting periods are extended from late summer to early spring (the coldest period of the year), and in all cases out of the breeding season. Hunting targets adults and juveniles (the latter in small game species), both males and females, which may have eaten natural or supplemented food depending on the location, as in many parts of Europe (especially the Mediterranean basin), there are extended periods of drought.

The different types of hunting may affect to meat quality. For example, driven hunting using dogs induce high levels of stress compared to stalking, in which the animals are often shot while being motionless. Other factors, such as hunting at day or night, with a rifle or a bow, may affect meat quality through different stress levels [13]. High stress levels before the animal being shot/captured, cause a higher consumption of muscle glycogen before the death of the animal, resulting in less *post-mortem* lactic acid generation from depleted glycogen reservoir [14], and affecting the pH meat value. An appropriate decrease on pH, and a final value ranging 5.5–5.7 [15], allows the meat to reach the desired color, tenderness, water retention capacity and other technological characteristics for transformation into meat products.

2.2 *Post-mortem* factors

After the death of the animal, the process to obtain meat in the field and the characteristics of the cutting plant varies depending on the location, as not all European countries have standard operating procedures. In fact, there are significant differences among countries in dressing the meat, the period of time from death to meat refrigeration, refrigeration temperatures, the type of vehicles used to transport the meat to the cutting plant/slaughterhouse, the maturation process which allows reaching the desired texture and flavor, and also the packing and storing system. All these factors affect the physicochemical, microbiological and sensory quality of game meat.

With regard to small game, after being shot, animals are usually kept unskinned and uneviscerated, and are hung in a cool, dry place until transportation to a game dealer or cutting plant. However, the majority of small game is generally kept by the hunters for personal consumption, and only a small proportion is sold.

In large game species, the evisceration is carried out by the hunter or the staff of the hunting ground. The interval from death to evisceration varies and little attention is paid to the environmental conditions in which hunting takes place, which may seriously undermine meat quality. The interval and the prevailing conditions between hunting and transport of the eviscerated carcass (previously approved by a veterinary inspection), to the cutting plant in refrigerated trucks also varies considerably. Finally, cutting plant managers often overlook the optimum conditions for carcass aging for a given species, which currently takes place over a wide range of temperatures and times, using different carcass ripening methods (mainly skinned *vs.* unskinned) [16].

The implementation of standardized procedures in the field and cutting plant, are needed to ensure safety, quality and traceability of game meat.

3. Nutritional composition of wild game meat

The macronutrients, micronutrients, cholesterol content and fatty acid profile for the meat from the most important hunted species in Central and Mediterranean European countries are described below. Existing literature from areas where hunting is very popular has been reviewed. Those areas are located at the countries: Italy, Spain, Portugal, Greek, Poland, Hungary, Romania, Croatia, Germany, Slovenia, Czech Republic, Slovak Republic and Republic of Lithuania. Only studies dealing with wild game (no farmed) have been cited, including animals harvested from late summer to spring, adults and juveniles, males and females.

3.1 Proximate composition

In **Table 1**, the proximate composition and energy content of meat from large wild game species is shown: wild red deer (*Cervus elaphus*), fallow deer (*Dama dama*), wild boar (*Sus scrofa*), and small game: brown hare (*Lepus europaeus*) and wild rabbit (*Oryctolagus cuniculus*), which have subject to a higher number of studies in Europe. The energy content has been calculated using the protein and fat content, following Regulation (EU) No 1169/2011. **Table 1** shows range of mean values or mean value when only one study available, found for the most studied parts of the carcass, loin and legs.

3.1.1 Large wild game

The most studied part of the carcass of red deer and fallow deer is the loin, both the whole *longissimus thoracis et lumborum* muscle (sometimes referred as *longissimus dorsi*) or its two parts (*longissimus thoracis and longissimus lumborum* muscles), with research also studying the legs (*semimembranosus, semitendinosus and triceps brachii* muscles). Generally speaking, the composition of both parts is similar, hence deer meat (red and fallow deer) has a high protein content ranging 20.5–22.8%, which confirms that is a relatively rich source of proteins and aminoacids [11]; and

		Moisture (g)	Protein (g)	Fat (g)	Ash (g)	Energy (kcal)
Red deer [17–21]	Loin	75.22–77.11	21.41–22.20	0.10–0.96	1.10–1.34	90–98
	Legs	77.90	20.50–21.50	1.00–1.72	0.98	91–101
Fallow deer [22–24]	Loin	74.30–75.40	21.80–22.80	0.30–0.89	1.07–1.10	91–96
	Legs	74.70–75.50	21.90–22.60	0.41–0.81	1.01	94–95
Wild boar [12, 25–32]	Loin	70.50–74.72	21.24–25.87	0.69–2.80	1.03–1.26	101–117
	Legs	71.28–76.74	19.71–23.73	1.30–2.80	0.88–1.05	95–111
Hare [33–36]	Loin	72.48–75.15	21.53–25.30	1.23–2.73	0.98–1.27	100–112
	Legs	74.35–75.43	20.28–21.58	1.79–3.23	0.97–1.27	102–110
Wild rabbit [37]	Loin	75.35	22.05	1.55	1.00	102
	Legs	75.08–75.55	20.28–21.43	2.26–3.48	1.02–1.09	106–113

Table 1.
Range of mean values (or mean value when only one study available) of proximate composition and energy content (per 100 g) of game meat from wild species hunted in Europe.

a low fat content between 0.1–1.7%. The loin is the part with less fat content (<1%). It is well documented that fat content of game meat as a high variability as it may be affected by several factors, including climate [22, 38], sex and age [20–22], feeding regime and season [22, 23, 39], physiological and sexual status [40], and hunting period [41].

In the wild, fat content is mostly influenced by seasonal variations, nutritional and sexual status, and available vegetation. Deer species, especially those living in the wild, are known to have seasonal body condition changes, which allows speculation about seasonal variation of meat composition. Usually, body condition is at its highest at the beginning of winter and at its lowest at the beginning of spring [28], and consequently the fat content is higher in winter compared to autumn and summer [22, 41]. But this variation is influenced by the species, winter severity and population density [28, 29]. In addition, in several deer species a decrease in body weight has been recorded for adult males during the rutting season [29], while for adult females the recovery of body condition during summer may be limited by the cost of lactation [29]. The meat energetic value (calculated per 100 g) ranges from 90 to 101 kcal in red deer, and 91.5–96 kcal in fallow deer. These values are lower than beef, pork, lamb or poultry meat, whose energetic value range from 114 to 231 kcal/100 g of muscle tissue [30] owing to the higher fat content of these meats. Ash content is very similar for both deer species and in all muscles (range of mean values from 1 to 1.3%).

When comparing meat from wild and farmed deer, there are contradictory results; higher fat content has been found in wild fallow deer compared to farmed, and similar or slightly higher values have been found in farmed red deer compared to wild ones [11].

The wild boar is a generalist omnivore, which is able to eat a wide range of food items, hence its diet varies according to resources available. It has a broad geographic distribution that include arid zones (semi deserts), wetlands, high mountain environments, forest and farmland ecosystems [25], which ultimately affect to the meat chemical composition. The proximate composition of loin (*longissimus thoracis et lumborum* and *longissimus thoracis*) and legs (*semimembranosus* and *teres major*) from European wild boar shows a high protein content (21.2–25.9%) and a low fat content (0.7–2.8%). The influence of some factors on wild boar meat has been studied, and the most significant are: (i) the available food in the hunting grounds which finally determines meat composition, outstanding fat and protein content. In this way, a higher fat content has been found in wild boar meat from fenced hunting grounds, in areas where cereals are supplemented and in cultivated forest compared to areas/hunting grounds where only natural food is available [25, 31, 32, 42–44]; (ii) age, as when compared to juveniles, meat from adults has higher content of fat and protein, and lower water content [33, 43, 44]; (iii) the carcass weight, as there is a positive correlation between the protein content and the carcass weight [34]; (iv) the hunting period, because it has been demonstrated that the highest content of intramuscular fat is reached in wild boar harvested in winter compared to spring and summer, probably owing to the higher amount of available food from cultivated crops [35]. Interestingly enough, sex has no significant effect on meat proximate composition [31], though a study [44] found that younger females generally showed a higher content of protein than males, and another one observed a higher content of protein in meat from males [12]. Among the different components studied, the ash content is the less influenced and more constant [44] with mean values ranging 0.9–1.3%.

When comparing the macronutrients content of wild boar meat to pork, the main difference is the lower fat and higher protein content of wild boar meat [25]. Thus, the mean values in pork loin are 4.6% for intramuscular fat and 21.4% for protein content [12, 25, 26, 34, 42, 43]. One study has reported that the mean fat content of wild boar meat was 5.3%, being similar to pork [23].

3.1.2 Small game

Wild rabbit and hares are herbivorous consuming a great variety of plants, including herbs, grains, and fruits. The type and abundance of these plants vary depending on the season and from one area to another, which may cause large variation in the composition of their meat [36]. There are very few data available regarding the nutritional composition of wild hare and rabbit in Europe; in fact, only one study has been found investigating wild rabbit meat harvested in Spain [37]. The proximate composition has been evaluated in different pieces such as loin (*longissimus thoracis et lumborum*) and legs (*semimembranosus* and *triceps brachii*). For both rabbits and hares, legs have a slightly higher fat content and lower protein content compared to loin, showing an overall a high protein content (20.3–25.3%) and low fat content (1.2–3.5%), resulting on an mean energetic value of 100–113 kcal/100 g. Although the meat composition of both lagomorphs harvested in Europe is similar, more research is needed to confirm this point.

Considering the variability for each of the parameters mentioned above, it is clear that the available food where animals are harvested has a significant influence on meat composition. On the other hand, some studies have found that several factors influence protein and fat content, and consequently energetic value, the most important being: (i) sex, as meat from females has a higher fat content and lower protein content [45, 46]; (ii) the hunting period, with higher fat and protein content in winter compared to spring [47]; (iii) muscle type, with a higher fat content in the legs compared to the loin [37, 46].

Several studies have compared meat from hares and domestic rabbit [45, 48], showing that domestic rabbit have a higher fat content (2.6–3.0%) and a lower protein value (21.5–22.2%). When the same diet is provided to farmed brown hares and domestic rabbits, a higher protein content is found in hare meat [48].

The comparison of the proximate composition of meat from wild and farmed brown hares, has found similar or slightly higher fat and mineral content in farmed hares [48–50].

Although meat from wild rabbit and hare provides a balanced content of macronutrient, these meats are not as popular as the consumption of other domestic meats, because of their dark red color, characteristic strong flavor, compact texture and dryness [51].

No complete studies have been found on the chemical composition of game birds, so more research is needed for these species.

In summary, the main characteristics of proximate composition of game meat from wild species are: low fat content and caloric value, together with high protein content, which may contribute to a healthy diet and a good "marketing strategy" [52].

3.2 Fatty acids composition

It is widely accepted that both the amount and the structure of fatty acids (FA) play a major role in human health. Moreover, the FA composition is more relevant than fat content for human nutrition. There are four inter-related factors that have important health implications: (i) the total fat content; (ii) the distribution of specific FA; (iii) the ratio of polyunsaturated fatty acids and saturated fatty acids (PUFA/SFA); (iv) the *n*-6/*n*-3 FA ratio [53]. Each of these factors has been shown to influence the development of coronary diseases.

According to the health authorities it is recommended to decrease the intake of fat, SFA, *trans*-FA and the *n*-6/*n*-3 ratio, and to increase the intake of MUFA (monounsaturated fatty acids) and PUFA [54]. SFA are regarded as the main cause of cardiovascular diseases as they increase blood pressure and the concentration of the "bad" LDL (low-density lipoproteins) serum cholesterol concentrations, while MUFA and PUFA decrease the concentration of the "bad" LDL cholesterol and increase the concentration of the "good" HDL (high-density lipoproteins) cholesterol which results in reducing the risk of heart diseases and atherosclerosis [55]. The *n*-3 PUFA decreases serum triacylglycerols, while lowering thrombic tendencies and reducing the incidence of ventricular arrhythmias [56]. Today, the *n*-6/*n*-3 ratio in human diets is often over 10 (the recommendation is <4.0), indicating the deficiency of beneficial *n*-3 FA in our diets [52]. On the other hand, the PUFA/SFA ratio recommended by the British Department of Health [57] is a minimum of 0.40 to contribute to a reduction in the risk in coronary diseases in humans.

As said before, game meat is lean, with a low content of intramuscular lipids, which are primarily composed of triglycerides (neutral lipids) and phospholipids (polar lipids) with polar lipids being the less saturated. The intramuscular fat composition is related to the fat amount, because a higher fat content results on changes in FA, with a higher content of neutral lipids which tend to be more saturated [58].

In **Table 2** a review of the range of mean values found in the available literature for the total content of SFA, MFA, PUFA, *n*-6 and *n*-3 FA, are shown, and also the PUFA/SFA and the *n*-6/*n*-3 ratios for the intramuscular fat of different muscles of loin (*longissimus thoracis et lumborum*, *longissimus thoracis* and *longissimus lumborum*) and legs (*semimembranosus, semitendinosus, psoas major* and *triceps brachii*).

3.2.1 Large wild game

The range of mean values for the total amount of FA and their ratios found in wild game meat harvested in Europe are quite wide. In general, in wild red deer meat the most important are SFA, closely followed by PUFA, whereas fallow deer shows a higher amount of SFA followed by MUFA. However, only one study is available for fallow deer [58] harvested in Poland hence there is lack of studies. Considering

	Red deer [17, 19, 59]		Fallow deer [24]	Wild boar [26, 42, 43, 60–64]		Hare [46, 48]	Wild rabbit [36]
	Loin	**Legs**	**Loin**	**Loin**	**Legs**	**Legs**	**Loin**
SFA	30.4–38.2	35.6–43.3	55.4–63.1	31.6–44.7	28.5–40.1	33.7–35.1	28.8
MUFA	15.3–22.7	22.1–37.2	25.7–29.4	30.2–46.8	27.8–50.4	17.5–21.8	13.6
PUFA	37.6–50.1	25.5–37.4	11.2–14.9	17.3–30.5	14.7–25.5	41.5–45.5	39.7
PUFA/ SFA		0.44–1.09	0.18–0.27	0.38–0.84	0.52–0.62	1.20–1.89	1.38
***n*-6**	30.2–38.2	18.7–29.3		15.7–28.0	17.8–24.0	39.3–41.1	28.8
***n*-3**	7.2–11.7	4.6–8.1		0.9–5.7	1.4–6.2	2.0–4.4	9.2
***n*-6/*n*-3**	3.5–4.3	2.6–4.8		3.1–22.3	3.0–17.6	14.2–25.1	3.8

Table 2.
Range of mean values (or mean value when only one study available) of fatty acids profile (g/100 g total fatty acids) of intramuscular fat from game meat from wild species hunted in Europe. SFA: Saturated fatty acids; MUFA: Monounsaturated fatty acids; PUFA: Polyunsaturated fatty acids.

all studies conducted on wild red deer, the *n*-6/*n*-3 ratio has shown values lower or close to the recommended maximum value of 4.0 [17, 19, 59] and PUFA/SFA values slightly higher than 0.40, from 0.44–0.49 [59] or slightly higher, from 0.63–1.09 [17], which is healthier. For fallow deer, this ratio was lower owing to the higher SFA content and lower PUFA content. Wild boar shows a higher amount of MUFA compared to ruminants, as shown in the majority of studies [42, 43, 59–63], and a higher *n*-6/*n*-3 ratio, between 6.9 and 17.0 which is 2 to 4 times above the maximum nutritional recommended ratio of 4.0, hence less favorable. Only a study carried out in Poland, shows lower values in wild boar loin and legs, ranging between 3.0–4.7 [26]. The majority of studies on wild boars harvested in Europe show PUFA/SFA ratios above the minimum ratio of 0.40 recommended to reduce the risk in coronary diseases in humans [60, 65].

Regarding to the FA in large game, all studies show that the most abundant are palmitic (C16:0), stearic (C18:0) and oleic (C18:1), as shown for domestic animals [66]. In general, game species contain high amounts of the long chained (C > 20) unsaturated fatty acids [67], which are healthful [68]. It is important to highlight the high amounts of arachidonic acid (C20:4 *n*-6), which can be explained by the phospholipidic origin of this fatty acid and the fact that wild animals meat is leaner than that of domestic animals [69].

In wild red and fallow deer, the FA composition is mainly influenced by the food composition despite the biohydrogenation of PUFA in the rumen. The intake of pasture produces a more favorable *n*-6/*n*-3 ratio compared to feed [68–70], which is attributed to the higher α-linolenic acid (ALA, C18:3 *n*-3) content in the grass, while the linoleic acid (C18:2 *n*-6) can be found in feed, and both are present primarily in the phospholipid fraction of deer intramuscular fat [17, 70]. Thus, it can be said that feed provided to farmed deer increases the total fat and SFA content, and decreases PUFA content, especially *n*-3 PUFA [53, 71–73]. Wild deer meat is generally speaking, a relatively good source of *n*-3 PUFA due to the presence of ALA in pasture. The high values of ALA found in red deer and fallow deer meat may be the consequence of the very low content of intramuscular fat, as well as the fact that both species are grazers.

Together with food, other factors may affect the FA profile as age, sex, area of harvesting, among others [74]. For example, meat from stags show a higher PUFA concentration than hinds in both red deer and fallow deer [58, 59, 75], and consequently the PUFA/SFA ratio is higher in the stags' meat. Regarding to the effect of age, younger animals show a higher PUFA content than older ones [19], probably due to higher intramuscular fat amounts in older animals and different relative contents of triacylglycerol and phospholipid fractions in muscle lipids. Meat from older animals shows higher n-6/n-3 ratio (> 4.0) [19], hence less healthy.

The comparison of meat from wild deer with other red meats from domestic animals, reveals a high content of PUFA, being deer meat particularly rich in C18:2 *n-6*, C18:3 *n-3* and C20:4 *n-6*. In this way, the content of C18:3 *n*-3 (ALA) in venison is markedly higher than in cattle breeds [70, 76–78]. On the one hand, red deer meat has a more favorable fat composition compared to grain-fed beef, showing a lower amount of SFA (2–3 times lower), a higher content in long-chain *n*-3 PUFA and healthier PUFA/SAF and *n*-6/*n*-3 ratios [53]. The practice of adding C18:2 *n*-6 to grain provided to cattle reduces the concentrations of *n*-3 PUFA, while simultaneously increases the concentrations of *n*-6 PUFA [53]. On the other hand, in deer meat conjugated linoleic acid, CLA (C18:2 *cis*–9 *trans*–11 isomer), has been found, an element which has been shown to be a cancer inhibitor that provides significant protection according to several models of carcinogenesis, especially in diets with a high intake of dairy products and beef, the highest known food sources. CLA may also reduce the progression of atherosclerosis via its strong antioxidant property [79], though deer meat has a lower CLA amount compared to meat from domestic ruminants [59].

In contrast to ruminants, double bonds of FA are not hydrogenated during digestion in wild boar. Similarly to other monogastric animals, the FA composition of wild boar's meat is strongly related to diet [25], as the FA profile varies depending on whether the animals eat only natural resources or are supplemented, the latter a quite widespread practice conducted by managers during harsh winter periods. In the wild, wild boars eat a great variety of indigenous plants, including grain, seeds, roots, fruits, insects, earthworms, slugs and also small mammals and carrion, though the bulk of food consumed consists of plant material [80]. Higher amounts of SFA have been found in wild boars supplemented with feed compared to those not supplemented [31], in which a higher content of *n*-3 was found [81]. In addition, the habitat influences the intramuscular FA, so wild boar meat from forest dominated habitats has a higher PUFA *n*-3 content and lower *n*-6/*n*-3 ratio compared to meat from wild boars occurring in farmland habitats [43]. It has been confirmed that the FA more affected by wild boar diet are C18:1 and C18:2 *n*-6, though SFA as C16:0 and C18:0, do not show significant variations [25]. The higher proportions of CLA in wild boar meat compared to pork [82] could be explained by the intestinal bacterial flora biosynthesis. As expected, the hunting area also influences the FA profile owing to the diversity of food items [42].

Together with the diet, it has been suggested that the hunting period has a strong influence on the FA profile. In autumn (October–November), research has found higher proportions of PUFA, lower of MUFA, more favorable values of PUFA/SFA ratio and less favorable *n*-6/*n*-3 ratio, when compared to winter (December–January) [62]. Also, in meat from wild boar hunted during winter, lower proportions of SFA and higher of PUFA, and PUFA/SFA and *n*-6/*n*-3 PUFA ratios have been found compared to summer [64].

With regard to the effect of age, conflicting results have been found, which could be attributed to the differences on age classes used among studies [62, 65]. Only small differences were detected for sex [62, 64], referring a higher total proportion of SFA in males compared to females [65]. A further influence has been found in the carcass weight, hence SFA increased with weight and PUFA proportion increased with decreased carcass weight [26].

If we look at the intramuscular FA profile in wild boars and pigs, in general higher amounts of PUFA have been found in wild boar but also a less favorable *n*-6/*n*-3 ratio [63, 83].

3.2.2 Small game

The FA profile of hares and rabbits is slightly different than the one shown by wild deer and wild boars, owing to the higher amount of C20:4 *n*-6, that increases the proportion of PUFA and decreases the proportion of MUFA when compared to large game species. The predominant FA in hare and wild rabbit are C18:2 *n*-6, 16:0, 18:1, 20:4 *n*-6 and 18:0 [36, 48], but hares have a higher *n*-6/*n*-3 ratio compared to rabbits. In any case, the fatty acids profile in hares and rabbits have been poorly studied, as only three studies have been conducted in Europe [36, 46, 48].

The FA composition is influenced by seasonal and territorial variations on vegetation. So, an optimal FA profile has been shown particularly during early spring, when fresh spontaneous vegetation is available, which increases the amount of PUFA, particularly *n*-3, and lower SFA and MUFA concentration, as demonstrated in wild rabbits hunted in Greece [36]. Additionally, there are differences on some FA depending on sex [46].

Higher amounts of C18:2, C18:3 *n*-3 and C20:4 *n*-4 have been found in hares and wild rabbits compared to livestock meats. In farmed hares and reared rabbits eating the same food, higher proportions of PUFA have been found in hares, particularly *n*-3, and lower proportions of MUFA and SFA [48].

3.2.3 Fatty acid profile and organoleptic characteristics

The FA profile not only has implications for human health, but also on meat quality, such as texture, shelf life, flavor and odor, which all affect sensory properties of deer meat [13]. Due to the different melting points, individual FA have important but diverse effects on the firmness or softness of the fat and meat. The effect of FA on meat flavor is due to the production of volatile lipid oxidation products during heat treatment or while the production of cured meat, especially unsaturated FA [70]. Natural grazing is a source of PUFA and it is considered as an important contributor for the development of "wild", "gamey "and "grassy" flavor in meat [72, 84, 85]. However, a high content of PUFA may negatively affect the oxidation stability and other technological parameters of meat and its products [52].

3.3 Cholesterol

Cholesterol is an essential element for body function, as it is part of the cell membrane structure and it is necessary for the sexual and adrenal hormone production. Around two thirds of cholesterol are synthesized in the liver and the remaining proportion is taken from the diet, and the cholesterol level is self-controlled. Higher serum cholesterol levels increase the risk of cardiovascular disease, though the intake reduction of food rich in cholesterol as a way to reduce this problem, has been questioned [86].

Available literature about wile game meat, shows cholesterol content only in wild red deer and wild boar, specifically in *longissimus thoracis et lumborum, semimembranosus, psoas major, semitendinosus and triceps brachii* muscles (**Table 3**). In those studies, the range of mean values are quite wide, which could explain by the differences in the method used for cholesterol determination, not

Red deer [17, 19, 59]		Wild boar [60, 87]	
Loin	**Legs**	**Loin**	**Legs**
45.3–52.8	55.2–94.6	34.4	20.9–56.9

Table 3.
Range of mean values (or mean value when only one study available) of cholesterol content (mg/100 g) of game meat from wild species hunted in Europe.

excluding the effects of diet, age and the muscle studied [17, 88]. The age influence is not clear yet [17], as lower cholesterol values in meat from younger red deer were found [59] and the opposite result was found in a different study [19]. As found for the total fat content, the amount of cholesterol is also influenced by the muscle, as a lower amount is found in the loin compared to the legs (**Table 3**).

The cholesterol amount in red deer meat has been compared to beef using the same technique at the same laboratory, finding that the amount found in deer is at the same level of the one found in beef loin and legs [89]. Depending on the muscle, the values found in wild boar meat (20.9–56.9 mg/100 g) are lower or similar to the ones found in pork, which are close to 45.3–80.0 mg/100 g [60]. Generally speaking, the amount of cholesterol in game meat are similar to the ones found in lamb, chicken, pork or beef meat [11]. Contradictory results have been found when comparing the cholesterol amount between wild and farmed deer, with wild red deer showing lower values and wild fallow deer showing higher values compared to farmed deer [11].

3.4 Minerals

Macro- and micro-minerals serve many functions in the human organism. They are the building material of bones, teeth, skin and hair and are fundamental components for metabolic processes, maintenance of acid–base equilibrium and regulation of water and electrolyte metabolism [90]. In addition, minerals contained in meat, in comparison with those present in plants, are more easily absorbed and, therefore, more beneficial to the human organism [91].

Table 4 shows the concentrations of macro and micro-elements, found in loin (*longissimus thoracis et lumborum, longissimus thoracis and longissimus lumborum*) and legs (*semimembranosus*) in wild red deer, fallow deer and wild boar. In despite of differences among species, the most abundant macro-minerals in the meat of all species are K, followed by P and Na. On the other hand, the most abundant micro-minerals are Fe and Zn, followed by Cu. The individual values for each species show wide ranges, as the studies have been conducted in different countries and areas, which results on different mineral availability. The mineral content of game meat, especially the micro-minerals, is closely linked to the availability in the environment, as soil micro-elements influence the vegetation content which is eaten by wild game [11, 12, 20, 23]. Additionally, other factors may affect mineral concentration in muscles for a given species, such as different physical activity, muscle fiber type composition, date of hunting, age, sex and other environmental factors [12, 19, 20, 74]. To date, it is known that age does not influence significantly the mineral profile, and only differences in Na have been found in red deer [19] and differences in Fe for wild boar, in which higher amounts are found in older animals [12]. Neither sex has a significant influence in wild board [92], though in red deer higher amounts of Na y Zn were found in stags [20] and in wild boar higher amounts of Cu were found in females [93]. Further research is needed to fully understand the factors influencing the mineral profile.

	Red deer [19–21, 23]		Fallow deer [23]		Wild boar [12, 23, 25, 92]	
	Loin	Legs	Loin	Legs	Loin	Legs
Macro-minerals						
Ca (mg)	6.0–25.1	9.1	21.0	9.0	14.8–20.8	10.1–11.8
P (mg)	166.1–289	187.9	226.9	216.8	202.3	214.3–223.2
Mg (mg)	17.5–38	18.6	21.4	21.3	20.1–55.7	19.6–26.9
K (mg)	217–326.9	282.9	302.0	303.3	299.7–1123.9	324.3–328.7
Na (mg)	45.7–121	83.5	63.5	76.2	69.4–157.2	60.9–68.1
Micro-minerals						
Cu (mg)	0.17–0.93	0.17	0.18	0.21	0.15–0.58	0.13–0.18
Zn (mg)	1.36–7.05	4.86	2.90	3.41	2.3–11.7	4.1–4.5
Fe (mg)	4.17–5.36	4.02	1.66	3.52	1.9–8.2	2.9–3.3
Mn (μg)	17–27	21.8	49.1	48.2	36.8–19.1	32.2–64.0
Se (μg)	3.7				13	

Table 4.
Range of mean values (or mean value when only one study available) of minerals content (per 100 g) of game meat from wild species hunted in Europe.

In general, game meat has higher amounts of micro-minerals than beef and pork, and such differences may be due to genetic background and environmental factors [23]. Muscles of wild ruminants have shown higher content of P, but less K and Na than corresponding bovine muscles [23]. In wild boar meat, higher content of Ca, P, Cu, Fe and Zn has been found than in pork, but lower content of Na and Mg, with conflicting results with regard to K [23, 92]. A 2-fold higher content of Fe have been found in wild boar [12].

3.5 Vitamins

Vitamins are essential elements for human nutrition as they control metabolic pathways of macronutrients, act in enzymatic reactions and take part in many physiological functions, with some vitamins having antioxidant properties [94].

Not many studies have dealt with the vitamin content of game meat, and available research focuses on vitamin E (tocopherols y tocotrienols) owing to the antioxidant properties which are key on the preservation of meat quality through the retardation of lipid oxidation and color loss. In **Table 5** it is shown the concentrations of B-group vitamins and some fat-soluble vitamins: retinol and vitamin E homologs which are at high levels, α- and γ-tocopherols, in red deer and wild boar meat.

In red deer, only one study has been conducted in B-group vitamins from loin (*longissimus thoracis*) [20], and when compared to lean meat from pork, poultry, lamb and beef [95], deer loin shows higher contents of thiamine (2-fold higher than poultry and lamb, and 3-fold higher than beef), riboflavin (being approximately 3-fold higher), and especially vitamin B12 which is between 3 and 5 times higher than all meats mentioned. One study in red deer evaluated the levels of tocopherols and tocotrienols in *psoas major* muscle, finding higher values of α-tocopherols in hinds [59].

	Red deer [20, 59]		Wild boar [12, 43, 60]	
	Loin	Legs	Loin	Legs
B-group (per 100 g)				
B1 (thiamine, mg)	0.18			
B2 (riboflavin, mg)	0.40			
B3 (niacin, mg)	5.53			
B6 (pyridoxine, mg)	0.14			
B9 (folic acid, μg)	3.17			
B12 (cyanocobalamine, μg)	5.87			
Retinol (μg /g)			0.01–1.11	
Tocopherols (μg /g)				
α		5.24–6.46	0.52–2.6	15.5–19.2
γ		1.42–2.06	0.03–0.30	1.14–1.75

Table 5.
Range of mean values (or mean value when only one study available) of vitamins content of game meat from wild species hunted in Europe.

In wild boar meat concentrations of retinol and vitamin E homologs have been studied in *longissimus* and *psoas major* muscles. The location of the animals has a strong influence on the amount of vitamins [12, 43, 60], finding significant differences for retinol, α- and γ-tocopherol in studies conducted in Italy and Germany. A study conducted in wild boar harvested at farmland and forest areas [43], found higher amount of α-tocopherol in farmland areas, and higher amount of γ-tocopherol in forest areas owing to the differences on crops, pasture and acorns among habitats. Supplementation of vitamin E in pigs is a widespread practice, as it reduces oxidation, increasing meat shelf life. Pigs supplemented up to 700 mg/kg of feed, resulted in α-tocopherol content in *psoas major* muscle up to 15.1–16.3 μg/g of meat, values that are similar to the mean values found in wild boars [60].

4. Nutrition and health claims in wild red deer meat

A recent study conducted by Soriano et al. [20] has shown the nutrition and health claims from wild red deer meat according to the Regulation (EC) No 1924/2006 and the amended Commission Regulation (EU) No 1047/2012. These claims can be used legally in labeling, presentation and advertising of this meat in the market. In that study the nutritional quality of loin (*longissimus thoracis*) from 71 wild Iberian red deer (*Cervus elaphus hispanicus*) was analyzed, using both stags and hinds hunted in autumn and winter.

Regarding to the nutrition claims, in all samples the contribution of proteins to the total energetic value was at least 73%, the fat content was lower than 2 g/100 g and Na content was lower than 0.12 g/100 g, so the claims "high protein", "low fat" and "low sodium/salt" can be assumed. When looking at the minerals and vitamins contents, significant amounts of P, Fe, Cu, Zn, and vitamins B2 (riboflavin), B3 (niacin) and B12 (cyanocobalamin) were found, indicating that 100 g of deer loin provide at least 15% of the reference daily intake for adults of these micronutrients, so this meat can be labeled as a "source of P, Fe, Cu, riboflavin and niacin". For Zn and vitamin B12 deer loin contains at least twice the value of source, so could be labeled as "high in Zn and vitamin B12".

Together with the mentioned nutrition claims, permitted health claims listed in the Annex of the Commission Regulation (EU) No 432/2012, could be used for wild deer meat in relation to those minerals and vitamins found in significant amounts. Similarly, claims related with children's development could also be made (Regulation (EC) No 1924/2006; Commission Regulation (EU) No 432/2012); these claims are based on the bioactivity and presence of certain components that must be found in significant amounts, such as Fe and protein, since deer loin complies as being a source of Fe and protein.

5. European wild game meat production and its challenges

Production of wild game meat as alternative to others red meats, offer both economic opportunities and nutritional benefits but is still a niche market to develop due to both low demand and supply limitations.

In Europe, hunting contributes only a small part to the overall meat supply, but represents a sustainable source of meat. Hunting is a legal activity and when managed properly, may contribute to wildlife conservation and rural economy [10]. Most hunting practices fulfill animal welfare requirements, though guidelines for sustainable hunting should be promoted to minimize ethical concerns.

The way in which game meat is produced differs from that of farmed animals, and prompts taking specific measures along the food chain [96]. Despite differences in game species, *ante-mortem* conditions, hunting procedures and *post-mortem* handling, there are common requirements regarding to meat safety and quality. To strengthening the wild game meat sector, it is essential to ensure that the meat is safe for the consumer through official inspections. To date, European legislation is limited to a series of general regulations on the hygiene of foodstuffs (Regulations (EC) Nos. 852/2004, 853/2004, 854/2004, 882/2004 and 2075/2005) as adapted into the legislation of state members. This means that the good practices recommended in Europe must be observed across the whole production sector: hunting grounds' owners, hunters and cutting plants. These good practices are crucial, as training is needed by those involved in this matter (including hunters, trained staff and official veterinarians), serving as reference when game carcasses or food companies trading meat from wild game are inspected. However, there are still specific unresolved questions and challenges [96].

On the other hand, the use of lead (Pb) ammunition is still widespread in many hunting modalities and it is well known that fragments of Pb can be found after impacting on the body of the animal, both in the place of impact and other nearby parts. A common practice is to eliminate the place of impact during carcass dressing. In addition, trading companies have metal detectors, which prevent the distribution of game meat with the presence of ammunition remains. However, wild animals inadvertently ingest Pb residues present at the soil and surface water in some areas, causing the Pb to be also present in their viscera. Consequently, there is a certain risk of exposure to lead in the population, mainly hunters and their families [97]. In this sense, EFSA has studied the lead dietary exposure in the European population from meat and meat products [98], finding particularly high results for wild boar meat. However, food consumed in larger quantities, as grains and grain products, milk and dairy products, non-alcoholic beverages and vegetables and vegetable products, have the greatest impact on lead dietary exposure, not being this the case for wild game meat [98]. No adverse effects from a single high intake of Pb have been reported, though chronic toxicity from repeated intakes has a toxic effect on the body, mainly on the central nervous system (particularly on the developing brain of young children and fetuses). On the other hand, inorganic Pb compounds have been classified by the International Agency for Research on Cancer (IARC) as probably carcinogenic to humans [99].

It is then clear that meat from wild game has several challenges that should be tackled to reduce detrimental effect on consumers, mainly those related to the type of ammunition used and some aspects related to handling *post-mortem* in the field.

6. Conclusions

Generally speaking wild game meat from Central and Mediterranean European countries, compared to livestock meat shows: (i) a lower fat content (< 3 g/100 g for large and < 4 g/100 g for small wild game species), and therefore a lower energy content; (ii) a higher or similar protein content; (iii) a positive fatty-acids profile, showing a higher proportion of PUFA, especially *n*-3, and consequently more favorable PUFA/SFA ratio, and in the specific case of wild boar, a relative adequate content of conjugated linoleic acid (CLA), and in the case of wild ruminants (red and fallow deer), an optimal *n*-6/*n*-3 ratio; (iv) a higher minerals content, mainly micro-minerals as Zn and the bioavailable form of heam Fe. Wild game meat has also optimal amounts of B-group vitamins, as riboflavin, niacin and B12, and vitamin E, with antioxidant properties. Overall, wild game meat meets current demands from consumers owing to its sustainable production which guarantees animal welfare standards, and its nutritional quality that may contribute to a balance and healthy diet. Hence wild game meat is a good alternative to others red meats and it can diversify the European meat market.

Author details

Almudena Soriano[1]* and Carlos Sánchez-García[2]

1 Food Technology, IRICA (Regional Institute for Applied Scientific Research), University of Castilla-La Mancha, Ciudad Real, Spain

2 Fundación Artemisan, Ciudad Real, Spain

*Address all correspondence to: almudena.soriano@uclm.es

References

[1] Hoffman LC, Muller M, Schutte DW, Calitz FJ, Crafford K. Consumer expectations, perceptions and purchasing of South African game meat. South African Journal of Wildlife Research. 2005; 35: 33-42

[2] Demartini E, Vecchiato D, Gaviglio A, Tempesta T, Viganò R. Consumer preferences for red deer meat: A discrete choice analysis considering attitudes towards wild game meat and hunting. Meat Science. 2018; 146: 168-179. DOI: 10.1016/j.meatsci.2018.07.031

[3] Kwiecińska K, Kosicka-Gębska M, Gębski J, Gutkowska K. Prediction of the conditions for the consumption of game by Polish consumers. Meat Science. 2017; 131: 28-33. DOI: 10.1016/j.meatsci.2017.04.038

[4] Andreotti A, Borghesi F, Aradis A. Lead ammunition residues in the meat of hunted woodcock: A potential health risk to consumers. Italian Journal of Animal Science. 2016; 15: 22-29. DOI: 10.1080/1828051X.2016.1142360

[5] Sevillano J, Moreno-Ortega A, Amaro MA, Arenas A, Cámara-Martos F, Moreno-Rojas R. Game meat consumption by hunters and their relatives: a probabilistic approach. Food Additives & Contaminants: Part A. 2018; 35: 1739-1748. DOI: 10.1080/19440049.2018.1488183

[6] Marescotti ME, Caputo V, Demartini E, Gaviglio A. Discovering market segments for hunted wild game meat. Meat Science; 2019; 149: 163-176. DOI: 10.1016/j.meatsci.2018.11.019

[7] European Commission-Directorate General for Agriculture and Rural Development. Study report-minor meats. September 2014; LEI Wageningen UR.

[8] Gibier de Chasse-Chasseurs de France. 2020. Available from: http://chasseurdefrance.com/charte-gibier-de-chasse-chasseurs-de-france/ [Accessed: 2020-10-19]

[9] ASICCAZA. 2020. Available from: http://asicaza.org/ [Accessed: 2020-10-19]

[10] Sánchez-García C, Urda V, Lambarri M, Prieto I, Andueza A, Villanueva LF. Evaluation of the economics of sport hunting in Spain through regional surveys. International Journal of Environmental Studies. 2020; 1-15. DOI: 10.1080/00207233.2020.1759305

[11] Kudrnáčová E, Bartoň L, Bureš D, Hoffman LC. Carcass and meat characteristics from farm-raised and wild fallow deer (*Dama dama*) and red deer (*Cervus elaphus*): A review. Meat Science. 2018; 141: 9-27. DOI: 10.1016/j.meatsci.2018.02.020

[12] Dannenberger D, Nuernberg G, Nuernber K, Hagemann E. The effects of gender, age and region on macro and micronutrient contents and fatty acid profiles in the muscles of roe deer and wild boar in Mecklenburg-Western Pomerania (Germany). Meat Science. 2013; 94: 39-46. DOI: 10.1016/j.meatsci.2012.12.010

[13] Neethling J, Hoffman LC, Muller M. Factors influencing the flavour of game meat: A review. Meat Science. 2016; 113: 139-153. DOI: 10.1016/j.meatsci.2015.11.022

[14] Pollard JC, Stevenson-Barry JM, Littlejohn RP. Factors affecting behaviour, bruising and pH in a deer slaughter premises. Proceedings of the New Zealand Society of Animal Production. 1999; 59: 148-151.

[15] Wiklund E, Farouk M, Finstad G. Venison: meat from red deer (*Cervus*

elaphus) and reindeer (*Rangifer tarandus tarandus*). Animal Frontiers. 2014; 4: 55-61. DOI: 10.2527/af.2014-0034

[16] Soriano A, Montoro V, Vicente J, Sánchez-Migallón BF, Benítez S, Utrilla MC, García Ruiz A. Influence of evisceration time and carcass ageing conditions on wild venison quality. Preliminary study. Meat Science. 2016; 114: 130-136. DOI: 10.1016/j.meatsci.2015.12.021

[17] Polak T, Rajar A, Gašperlin L, Žlender B. Cholesterol concentration and fatty acid profile of red deer (*Cervus elaphus*) meat. Meat Science. 2008; 80: 864-869. DOI: 10.1016/j.meatsci.2008.04.005

[18] Daszkiewicz T, Janiszewski P, Wajda S. Quality characteristics of meat from wild red deer (*Cervus elaphus* L.) hinds and stags. Journal of Muscle Foods. 2009; 20: 428-448. DOI: 10.1111/j.1745-4573.2009.00159.x

[19] Lorenzo JM, Maggiolino A, Gallego L, Pateiro M, Pérez Serrano M, Domínguez R, García A, Landete-Castillejos T, De Palo P. Effect of age on nutritional properties of Iberian wild red deer meat. Journal of the Science of Food and Agriculture. 2019; 99: 1561-1567. DOI: 10.1002/jsfa.9334

[20] Soriano A, Murillo P, Perales M, Sánchez-García C, Murillo A, García Ruiz A. Nutritional quality of wild Iberian red deer (*Cervus elaphus hispanicus*) meat: effects of sex and hunting period. Meat Science. 2020; 168: 108189. DOI: 10.1016/j.meatsci.2020.108189

[21] Serrano MP, De Palo P, Maggiolino A, Pateiro M, Gallego L, Domínguez R, García A, Landete-Castillejos T, Lorenzo JM. Seasonal variations of carcass characteristics, meat quality and nutrition value in Iberian wild red deer. Spanish Journal of Agriculture Research. 2020; 18. DOI: 10.5424/sjar/2020184-16113

[22] Stanisz M, Skorupski M, Ślósarz P, Bykowska-Maciejewska M, Składanowska-Baryza J, Stańczak Ł, Krokowska-Paluszak M, Ludwiczak A. The seasonal variation in the quality of venison from wild fallow deer (*Dama dama*) – a pilot study. Meat Science. 2019; 150: 56-64. DOI: 10.1016/j.meatsci.2018.12.003

[23] Zomborszky Z, Szentmihály G, Sarudi I, Horn P, Szabó CS. Nutrient composition of muscles in deer and boar. Journal of Food Science. 1996; 61: 625-627. DOI: 10.1111/j.1365-2621.1996.tb13172.x

[24] Piaskowska N, Daszkiewicz T, Kubiak D, Janiszewski P. the effect of gender on meat (*longissimus lumborum* muscle) quality characteristics in the fallow deer (*Dama dama* L.). Italian Journal of Animal Science. 2015; 14: 389-393. DOI: 10.4081/ijas.2015.3845

[25] Sales J, Kotrbaet R. Meat from wild boar (*Sus scrofa* L.): A review. Meat Science. 2013; 94: 187-201. DOI: 10.1016/j.meatsci.2013.01.012

[26] Batorska M, Więcek J, Kunowska-Slósarz M, Puppel K, Slósarz J, Gołębiewski M, Kuczyńska B, Popczyk B, RekielV A, Balcerak M. The effect of carcass weight on chemical characteristics and fatty acid composition of *Longissimus dorsi* and *Semimembranosus* muscles of European wild boar (*Sus scrofa scrofa*) meat. Canadian Journal of Animal Science. 2018; 98: 557-564. DOI: 10.1139/cjas-2017-0090

[27] Stanisz M, Ludwiczak A, Składanowska-Baryza, AJ, Bykowska-Maciejewska M. The effect of age and ultimate pH value on selected quality traits of meat from wild boar. Canadian Journal of Animal Science.

2019; 99: 336-342. DOI: 10.1139/cjas-2018-0090

[28] Hjeljord O, Histøl T. Range-body mass interactions of a northern ungulate – a test of hypothesis. Oecologia. 1999; 119: 326-339. DOI: 10.1007/s004420050793

[29] Hewison AJM, Angibault JM, Boutin J, Bideau E, Vincent JP, Sempéré A. Annual variation in body composition of roe deer (*Capreolus capreolus*) in moderate environmental conditions. Canadian Journal of Zoology. 1996; 74: 245-253. DOI: 10.1139/z96-031

[30] Chizzolini R, Zanardi E, Dorigoni V, Ghidini S. Calorific value and cholesterol content of normal and low-fat meat and meat products. Trends in Food Science and Technology. 1999; 10: 119-128. DOI: 10.1016/S0924-2244(99)00034-5

[31] Skobrák EB, Bodnár K, Jónás EM, Gundel J, Jávor A. The comparison analysis of the main chemical composition parameters of wild boar meat and pork. Scientific Papers Animal Science and Biotechnologies. 2011; 44: 105-112

[32] Żochowska-Kujawska J, Lachowicz K, Sobczak M, Bienkiewicz G. Utility for production of massaged products of selected wild boar muscles originating from wetlands and an arable area. Meat Science. 2010; 85: 461-466. DOI: 10.1016/j.meatsci.2010.02.016

[33] Ludwiczak A, Składanowska-Baryza J, Stanisz M. Effect of Age and Sex on the Quality of Offal and Meat of the Wild Boar (*Sus scrofa*). Animals. 2020; 10: 660-670. DOI: 10.3390/ani10040660

[34] Kasprzyk A, Stadnik J, Stasiak D. Technological and nutritional properties of meat from female wild boars (*Sus scrofa scrofa* L.) of different carcass weights. Archives Animal Breeding. 2019; 62: 597-604. DOI: 10.5194/aab-62-597-2019

[35] Lachowicz K, Żochowska-Kuiawska J, Gaiowiecki L, Sobczak M, Kotowicz M, Żych A. Effects of wild boar meat of different seasons of shot: addition texture of finely ground model pork and beef sausages. Electronic Journal of Polish Agricultural Universities. 2008; 11: e11

[36] Papadomichelakis G, Zoidis E, Pappas AC, Hadjigeorgiour I. Seasonal variations in the fatty acid composition of Greek wild rabbit meat. Meat Science. 2017; 134: 158-162. DOI: 10.1016/j.meatsci.2017.08.001

[37] Cobos A, de la Hoz L, Cambero MI, Ordóñez JA. Chemical and fatty acid composition of meat from Spanish wild rabbits and hares. Zeitschrift für Lebensmittel-Untersuchung und-Forschung. 1995; 200: 182-185

[38] Dahlan I, Norfarizan Hanoon NA. Chemical composition, palatability and physical characteristics of venison from farmed deer. Animal Science Journal. 2008; 79: 498-503. DOI: 10.1111/j.1740-0929.2008.00555.x

[39] Suttie JM, Goodall ED, Pennie K, Kay RNB. Winter food restriction and summer compensation in red deer stags (*Cervus elaphus*). British Journal of Nutrition. 1983; 50: 737-747. DOI: 10.1079/BJN19830145

[40] Stevenson JM, Seman DL, Littlejohn RP. Seasonal variation in venison quality of mature farmed, red deer stags in New Zealand. Journal of Animal Science. 1992; 70: 1389-1396. DOI: 10.2527/1992.7051389x

[41] García Ruiz A, Mariscal C, Gonzalez Viñas MA, Soriano A. Influence of hunting-season stage and ripening conditions on microbiological,

physicochemical and sensory characteristics of venison (*Cervus elaphus*) chorizo sausages. Italian Journal of Food Science. 2010; 22: 386-394

[42] Amici A, Cifuni GF, Contò M, Esposito L, Failla S. Hunting area affects chemical and physical characteristics and fatty acid composition of wild boar (*Sus scrofa*) meat. Rendiconti Lincei. 2015; 26: S527-S534. DOI: 10.1007/s12210-015-0412-7

[43] Pedrazzoli M, Dal Bosco A, Castellini C, Ranucci D, Mattioli S, Pauselli M, Roscini V. Effect of age and feeding area on meat quality of wild boars. Italian Journal of Animal Science. 2017; 16: 353-362 DOI: 10.1080/1828051X.2017.1292114

[44] Tesařová S, Ježek F, Hulánková R, Plhal R, Drimaj J, Steinhauserová I, Bořilová G. The individual effect of different production systems, age and sex on the chemical composition of wild boar meat. Acta Veterinaria Brno. 2018; 87: 395-402. DOI: 10.2754/avb201887040395

[45] Mertin D, Slamečka J, Ondruška Ľ, Zaujec K, Jurčík R, Gašparík J. Comparison of meat quality between european brown hare and domestic rabbit. Slovak Journal of Animal Science. 2012; 45: 89-95

[46] Frunza G, Simeanu D, Pop C, Lazar R, Boisteanu PC, Stefan M. Contributions on the sensorial, physico-chemical and nutritional characterization of hare meat (*Lepus europaeus pallas*). Revista de Chimie. 2019; 70: 174-180. DOI: 10.37358/RC.19.1.6876

[47] Skrivanko M, Hadziosmanovic M, Cvrtila Z, Zdolec N, Filipovic I, Kozaeinski L, Florijanèic T, Boskovic I. The hygiene and quality of hare meat (*Lepus europaeus* Pallas) from Eastern Croatia. Archiv fur Lebensmittelhygiene. 2008; 59: 180-184

[48] Króliczewska B, Miśta D, Korzeniowska M, Pecka-Kiełb E, Zachwieja A. Comparative evaluation of the quality and fatty acid profile of meat from brown hares and domestic rabbits offered the same diet. Meat Science. 2018; 145: 292-299. DOI: 10.1016/j.meatsci.2018.07.002

[49] Vicenti A, Ragni M, di Summa A, Marsico G, Vonghia G. Influence of feeds and rearing system on the productive performances and the chemical and fatty acid composition of hare meat. Food Science and Technology International. 2003; 9: 279-283. DOI: 10.1177/108201303038106

[50] Trocino A, Birolo M, Dabbou S, Gratta F, Rigo N, Xiccato G. Effect of age and gender on carcass traits and meat quality of farmed brown hares. Animal. 2018; 12: 864-871. DOI: 10.1017/S1751731117002385

[51] Cambero MI, de la Hoz L, Sanz B, Ordóñez JA. Seasonal variations in lipid composition of Spanish wild rabbit (*Oryctolagus cuniculus*) meat. Journal of the Science of Food and Agriculture. 1991; 56: 315-362. DOI: 10.1002/jsfa.2740560311

[52] Poławska E, Cooper RG, Jóźwik A, Pomianowski J. Meat from alternative species-Nutritive and dietetic value, and its benefit for human health-A review. CyTA-Journal of Food. 2013; 11: 37-42. DOI: 10.1080/19476337.2012.680916

[53] Cordain L, Watkins BA, Florant GL, Kelher M, Rogers L, Li Y. Fatty acid analysis of wild ruminant tissues: evolutionary implications for reducing diet-related chronic disease. European Journal of Clinical Nutrition. 2002; 56: 181-191. DOI: 10.1038/sj/ejcn/1601307

[54] WHO, FAO. Diet, Nutrition and the Prevention of Chronic Diseases. Geneva: World Health Organization Technical Report Series; 2003. 916. 149 p

[55] Gerhard GT, Ahmann A, Meeuws K. Effect of a low-fat diet compared with those of a high-monounsaturated fat diet on body weight, plasma lipids and lipoproteins, and glycemic control in type 2 diabetes. The American Journal of Clinical Nutrition. 2004; 80: 668-673. DOI: 10.1093/ajcn/80.3.668

[56] Leaf A, Kang JX, Xiao YF, Billman GE. n-3 fatty acids in the prevention of cardiac arrhythmias. Lipids. 1999; 34: S187-S189

[57] British Department of Health. Nutritional aspects of cardiovascular disease. Report of the Cardiovascular Review Group of the Committee of Medical Aspects of Food Policy. London: HMSO; 1994. 46. 186 p

[58] Hunt MR, Legako JF, Dinh TTN, Garmyn AJ, O'Quinn TG, Corbin CH, Rathmann RJ, Brooks JC, Miller MF. Assessment of volatile compounds, neutral and polar lipid fatty acids of four beef muscles from USDA Choice and Select graded carcasses and their relationships with consumer palatability scores and intramuscular fat content. Meat Science. 2016; 116, 91-101. DOI: 10.1016/j.meatsci.2016.02.010

[59] Quaresma MAG, Trigo-Rodrigues I, Alves SP, Martins SIV, Barreto AS, Bessa RJB. Nutritional evaluation of the lipid fraction of Iberian red deer (*Cervus elaphus hispanicus*) tenderloin. Meat Science. 2012; 92: 519-524. DOI: 10.1016/j.meatsci.2012.05.021

[60] Quaresma MAG, Alves SP, Trigo-Rodrigues I, Pereira-Silva R, Santos N, Lemos JPC, Barreto AS, Bessa RJB. Nutritional evaluation of the lipid fraction of feral wild boar (*Sus scrofa scrofa*) meat. Meat Science. 2011; 89:457-461. DOI: 10.1016/j.meatsci.2011.05.005

[61] Gálik B, Šmehýl P, Gašparík J, Candrák J, Jahnátek A, Bíro D, Rolinec M, Juráček M, Šimko M. The effect of age on the fatty acids composition in wild boar (*Sus scrofa*) hunted in the southwest region of Slovakia. Acta Veterinaria Brno. 2018; 87: 85-90. DOI: 10.2754/avb201887010085

[62] Russo C, Balloni S, Altomonte I, Martini M, Nuvoloni R, Cecchi F, Pedonese F, Salari F, Marilia Sant'ana Da Silva A, Torracca B, Profumo A. Fatty acid and microbial profile of the meat (*longissimus dorsi* muscle) of wild boar (*Sus scropha scropha*) hunted in Tuscany. Italian Journal of Animal Science. 2017; 16: 1-8. DOI: 10.1080/1828051X.2016.1261006

[63] Stasiak K, Roślewska A, Stanek M, Jankowiak H, Cygan-Szczegielniak D, Bocian M. Comparison of the fatty acid profile in the meat of pigs and wild boars. Italian Journal of Food Science. 2018; 30: 707-714. DOI: 10.14674/IJFS-1187

[64] Razmaitė V, Šiukščius A. Seasonal variation in fatty acid composition of wild boar in Lithuania. Italian Journal of Animal Science. 2019; 18: 350-354. DOI: 10.1080/1828051X.2018.1530957

[65] Razmaitė V, Švirmickas GJ, Šiukščius A. Effect of weight, sex and hunting period on fatty acid composition of intramuscular and subcutaneous fat from wild boar. Italian Journal of Animal Science. 2012; 11: 174-179. DOI: 10.4081/ijas.2012.e32

[66] Dugan LR. Química de los tejidos animales. Parte 2 Grasas. In: Price JF, Schweigert BS, editors. Ciencia de la Carne y de los Productos Cárnicos. 2nd ed. Zaragoza: Acribia; 1994. p. 93-101

[67] Hoffman LC, Wiklund E. Game and venison-meat for the modern consumer. Meat Science. 2006; 74: 197-208. DOI: 10.1016/j.meatsci.2006.04.005

[68] Givens ID, Gibbs RA. Current intakes of EPA and DHA in European

populations and the potential of animal-derived foods to increase them. Proccedings of the Nutrition Society. 2008; 67: 272-280. DOI: 10.1017/S0029665108007167P

[69] Miller, GJ, Field RA, Riley ML, Williams JC. Lipids in wild ruminant animals and steers. Journal of Food Quality. 1986; 9: 331-343. DOI: 10.1111/j.1745-4557.1986.tb00802.x

[70] Wood JD, Richardson RI, Nute GR, Fisher AV, Campo MM, Kasapidou E, Sheard PR, Enser M. Effects of fatty acids on meat quality: A review. Meat Science. 2003; 66: 21-32. DOI: 10.1016/S0309-1740(03)00022-6

[71] Volpelli LA, Valusso R, Morgante M. Pittia P, Piasentier E. Meat quality in male fallow deer (*Dama dama*): Effects of age and supplementary feeding. Meat Science. 2003; 65: 555-562. DOI: 10.1016/S0309-1740(02)00248-6

[72] Wiklund E, Manley T, Littlejohn R, Stevenson-Barry J. Fatty acid composition and sensory quality of Musculus *longissimus* and carcass parameters in red deer (*Cervus elaphus*) grazed on natural pasture or fed a commercial feed mixture. Journal of the Science of Food and Agriculture. 2003; 83: 419-424. DOI: 10.1002/jsfa.1384

[73] Volpelli LA, Valusso R, Piasentier E. Carcass quality in male fallow deer (*Dama dama*): effects of age and supplementary feeding. Meat Science. 2002; 60: 427-432. DOI: 10.1016/S0309-1740(01)00156-5

[74] Hoffman LC, Kroucamp M, Manley M. Meat quality characteristics of springbok (*Antidorcas marsupialis*). 2: Chemical composition of springbok meat as influenced by age, gender and production region. Meat Science. 2007; 76:762-767. DOI: 10.1016/j.meatsci.2007.02.018

[75] Purchas RW, Triumf EC, Egelandsdal B. Quality characteristics and composition of the longissimus muscle in the short-loin from male and female farmed red deer in New Zealand. Meat Science. 2010; 86: 505-510. DOI: 10.1016/j.meatsci.2010.05.043

[76] Bureš D, Bartoň L, Kotrba R, Hakl J. Quality attributes and composition of meat from red deer (*Cervus elaphus*), fallow deer (*Dama dama*) and Aberdeen Angus and Holstein cattle (*Bos taurus*). Journal of the Science of Food and Agriculture. 2015; 95: 2299-2306. DOI: 10.1002/jsfa.6950

[77] Nuernberg K, Dannenberger D, Nuernberg G, Ender K, Voigt J, Scollan ND, Wood JD, Nute GR, Richardson, R. I. Effect of a grass-based and a concentrate feeding system on meat quality characteristics and fatty acid composition of longissimus muscle in different cattle breeds. Livestock Production Science. 2005; 94: 137-147. DOI: 10.1016/j.livprodsci.2004.11.036

[78] Realini, C. E., Duckett, S. K., Brito, G. W., Dalla Rizza, M., & De Mattos, D. Effect of pasture *vs.* concentrate feeding with or without antioxidants on carcass characteristics, fatty acid composition, and quality of Uruguayan beef. Meat Science. 2004; 66: 567-577. DOI: 10.1016/S0309-1740(03)00160-8

[79] Rudel LL. Atherosclerosis and conjugated linoleic acid. British Journal of Nutrition. 1999; 81:177-179. DOI: 10.1017/S0007114599000367

[80] Schley L, Roper TJ. Diet of wild boar Sus scrofa in Western Europe, with particular reference to consumption of agricultural crops. Mammal Review. 2003; 33: 43-56. DOI: 10.1046/j.1365-2907.2003.00010.x

[81] Marsico G, Rasulo A, Dimatteo S, Tarricone S, Pinto F, Ragni M. Pig, F1 (wild boar×pig) and wild boar meat quality. Italian Journal of Animal Science. 2007; 6: 701-703. DOI: 10.4081/ijas.2007.1s.701

[82] Chin SF, Liu W, Storkson JM, Ha YL, Pariza MW. Dietary sources of conjugated dienoic isomers of linoleic acid, a newly recognized class of anticarcinogens. Journal of Food Composition and Analysis. 1992; 5: 185-197. DOI: 10.1016/0889-1575(92)90037-K

[83] Barbani R, Bonilauri P, Sangiorgi E, Marliani G. Fatty acid composition of wild boar (*Sus scrofa scrofa*) meat compared to commercial hybrid and crossbreed Mora Romagnola swine. A preliminary study. Rivista Italiana delle Sostanze Grasse. 2020; 97: 7-14

[84] Finstad G, Wiklund E, Long K, Rincker PJ, Oliveira ACM, Bechtel PJ. Feeding soy or fish meal to Alaskan reindeer (*Rangifer tarandus tarandus*)-Effects on animal performance and meat quality. Rangifer. 2007; 27: 59-75. DOI: 10.7557/2.27.1.190

[85] Wiklund E, Johansson L, Malmfors G. Sensory meat quality, ultimate pH values, blood parameters and carcass characteristics in reindeer (*Rangifer tarandus tarandus L.*) grazed on natural pastures or fed a commercial feed mixture. Food Quality and Preference. 2003; 14: 573-581. DOI: 10.1016/S0950-3293(02)00151-9

[86] Nawar WW. Lipidos. In: Fennema OR, editor. Química de los Alimentos. 2nd ed. Zaragoza: Editorial Acribia; 1993. p. 157-274

[87] Skewes O, Morales R, Mendoza N, Smulders FJM, Paulsen P. Carcass and meat quality traits of wild boar (*Sus scrofa* s. L.) with 2n=36 karyotype compared to those of phenotypically similar crossbreeds (2n=37 and 2n=38) raised under the same farming conditions 2: Fatty acid profile and cholesterol. Meat Science. 2009; 83: 195-200. DOI: 10.1016/j.meatsci.2009.04.017

[88] Ramanzin M, Amici A, Casoli C, Esposito L, Lupi P, Marsico G, Mattiello S, Olivieri O, Ponzetta MP, Russo C, Marinucci MT. Meat from wild ungulates: ensuring quality and hygiene of an increasing resource. Italian Journal of Animal Science. 2010; 9: 318-331. DOI: 10.4081/ijas.2010.e61

[89] Alfaia CMM, Ribeiro VSS, Lourenço MRA, Quaresma MAG, Martins SIV, Portugal APV, Fontes CMGA, Bessa RJB, Castro MLF, Prates JAM. Fatty acid composition, conjugated linoleic acidisomers and cholesterol in beef from crossbred bullocks intensively produced andfrom Alentejana purebred bullocks reared according to Carnalentejana-PDO specifications. Meat Science. 2006; 72: 425-436. DOI: 10.1016/j.meatsci.2005.08.012

[90] Soetan KO, Olaiya CO, Oyewole OE. The importance of mineral elements for humans, domestic animals and plants: a review. African Journal of Food Science. 2010. 4:200-222. DOI: 10.5897/AJFS.9000287

[91] de Castro Cardoso Pereira PM, dos Reis Baltazar Vicente AF. Meat nutritional composition and nutritive role in the human diet. Meat Science. 2013; 93: 586-592. DOI: 10.1016/j.meatsci.2012.09.018

[92] Babicz M, Kasprzyk A. Comparative analysis of the mineral composition in the meat of wild boar and domestic pig. Italian Journal of Animal Science. 2019; 18: 1013-1020, DOI: 10.1080/1828051X.2019.1610337

[93] Roslewska A, Stanek M, Janicki B, Cygan-Szczegielniak D, Stasiak K, Buzala, M. Effect of sex on the content of elements in meat from wild boars (*Sus scrofa* L.) originating from the province of Podkarpacie (south-eastern Poland). Journal of Elementology. 2016. 21: 823-832. DOI: 10.5601/jelem.2015.20.2.943

[94] Badui S. Vitaminas y nutrimentos inorgánicos. In: Badui S, editor. Química

de los Alimentos. 4th ed. México: Pearson; 2006. p. 363-398

[95] Moreiras O, Carbajal A, Cabrera L, Cuadrado C, editors. Tablas de Composición de Alimentos. 15th ed. España: Ediciones Pirámide; 2008. p. 104-107.

[96] Winkelmayer R, Stangl PV, Paulsen P. Assurance of food safety along the game meat production chain: inspection of meat from wild game and education of official veterinarians and 'trained persons' in Austria. In: Paulsen P, Bauer A, Vodnansky M, Winkelmayer R, Smulders FJM, editors. Game Meat Hygiene in Focus: Microbiology, Epidemiology, Risk Analysis and Quality Assurance. 1st ed. The Netherlands: Wageningen Academic Publishers; 2011. p. 245-258

[97] AESAN (Agencia Española de Seguridad Alimentaria y Nutrición). 2012. Informe del Comité Científico de la Agencia Española de Seguridad Alimentaria y Nutrición (AESAN) sobre el riesgo asociado a la presencia de plomo en carne de caza silvestre en España. Available from: https://www.aetox.es/ [Accessed: 2021-01-10]

[98] EFSA (European Food Safety Authority). Lead dietary exposure in the European population. EFSA Journal. 2012; 10(7):2831 [59 pp.]. Available from: www.efsa.europa.eu/efsajournal/ [Accessed: 2021-01-10]. DOI:10.2903/j.efsa.2012.2831.

[99] IARC (International Agency for Research on Cancer). Inorganic and Organic Lead Compounds. In: IARC Monographs on the Evaluation of Carcinogenic Risks to Humans; 2006. p. 87

Chapter 6

Role of Biogenic Amines in Protein Foods Sensing: Myths and Evidence

Raffaela Biesuz and Lisa Rita Magnaghi

Abstract

Myriads of sensors have been proposed to signal the spoilage of a piece of meat. It is assumed and taken for granted that biogenic amines, BAs, harmful by-product indicating the last phase of degradation, must be present in the volatilome developed over the decaying meat. This chapter aims to clearly explain BAs' role in protein food spoilage: undoubtedly produced inside the meat, never present in the headspace, where sensors are displayed. The BAs presence in the headspace represents a sort of myth. It is plenty of evidence that BAs cannot be present in the volatilome over the meat. The BAs' volatility is pH-dependent. As shown by their protonation constants, the strongly buffered pH of proteinaceous food prevents their vapour phase transition. The chemical analyses made at the same degradation time, on the meat and the headspace above the meat, corroborate the real composition of the volatilome, demonstrating the BAs absence. The sensors here described, designed on volatilome evidence, succeed to follow the entire process, from the SAFE condition to the WARNING and the HAZARD. The final prototype works reliably on real protein foods (i.e. chicken, beef pork and fish), not enriched and stored at the home condition.

Keywords: smart labels, entire meat spoilage detection, naked-eye devices, BAs volatility, pH indicators, intelligent packaging

1. Introduction

In recent decades, the interest in systems able to detect food degradation, giving a quick response, simple to be read, and suitable for implementation in control systems and smart labels has increased continuously. [1–6] This interest is more valid in foods such as meat and fish. Despite the spread of vegan and vegetarian diets, their consumption has been steadily increasing worldwide for decades, also caused by the dramatic economic development of Eastern Country of the last forty years. The impressive trend is visibly depicted in the two graphs of **Figures 1** and **2**. [7] Unfortunately, the common feature of meat food is its intrinsically high perishability. All protein foods require proper processing, highly controlled environments, and storage in chilled conditions (cold rooms, refrigerators, and freezers), being very sensitive to interruptions in the cold chain. Furthermore, consumers pay increasing attention to their food quality and claim to have direct control over freshness.

IntechOpen

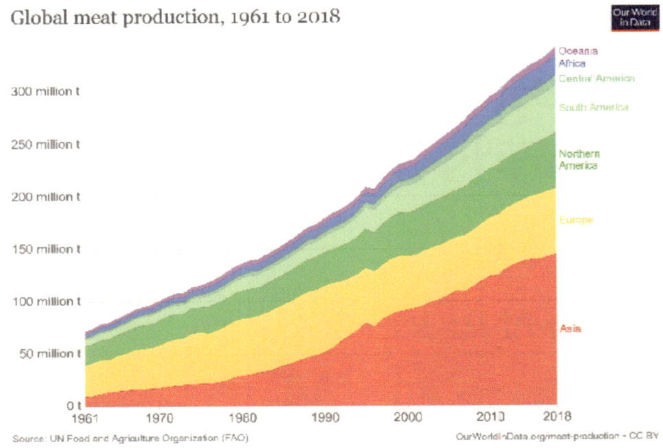

Figure 1.
Worldwide meat production from 1961 to 2018, as reported in reference [7].

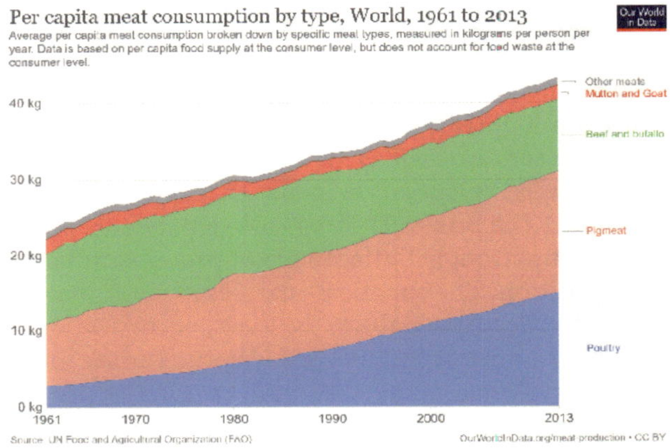

Figure 2.
Worldwide capita meat consumption from 1961 to 2013, as reported in reference [7].

Establish the quality of protein-based foods is anything but trivial. The first concern that makes the freshness assessment paramount is to reduce the risk of food poisoning, which is classified according to the responsible microorganism (bacteria, viruses, or parasites). The foodborne diseases could be the consequence of the consumption that has exceeded the expiration date. Significant contributions also come from lousy storage, adverse events along the supply chain, or the consumer's improper manipulation. Food poisoning remains a public health issue and still affects 23 million people a year only in Europe. [8] Secondly, but no less critical, freshness assessment is mandatory to reduce unjustified food waste. The expiration date is established conservatively, as a consequence of what assessed above. The ambiguity of food labels (*use by* date, *best before* date, *expiry* date, *sell by* date), is sometimes misleading. Consequently, too many people throw away food only because it has passed the expiration date, often sacrificing food that is still perfectly safe. According to FAO, the social cost of food waste is around $ 882 billion. Food waste is worth € 15.5 billion per year in Italy, about 1% of Italian GDP; € 3 billion comes from waste during the food chain, between production and distribution. [9] In this scenario, the foods' quality control systems include specific analytical measures that provide reliable and irrefutable data but are limited for control agencies, requiring long-time analyses, equipped laboratories, and trained personnel.

Instead, the possibility of knowing the quality of the single piece of meat/fish that was purchased in the local supermarket and forgotten in the bottom of the fridge must be based on simple systems of low cost and easy to understand. Consequently, there has been a great effort in developing devices to this purpose. [2] The largest category of prototypes, also in an advanced commercial step, is focused on collateral properties, indirectly related to the food freshness. Devices aimed to control the cold chain's maintenance, [3, 10] or the modified atmosphere's integrity for the food packed under CO_2 [11] are clear examples. Sophisticated control of the meat's colour and the atmosphere required apps or trained people, and are far to give instrument-less and straightforward responses. It is also true for all electronic noses or tongues. [12–14], or devices that require a sophisticated algorithm. [15] Undoubtedly, the colourimetric sensors are the best candidates to develop a device to assess the freshness. These sensors, placed in the tray over the meat, revel the volatilome change from safe to hazard conditions through a simple colour change. These devices are based on a colour code, developed as an on/off strategy or a change of colour; they intrinsically offer an easy interface to the consumer or untrained people, as supermarket staff. The literature is plenty of proposals exploited on this principle, mainly based on pH indicators. [16–25].

In principle, for each degradation step, the volatile by-products exhibit different acid–base behaviour, resulting in changes in the headspace's acidity over meat samples. For instance, in beef meat, the early post-mortem (within one hour of slaughter) pH is between 6.7-7.1 [26] which does not differ from that of other dead animals. Indeed, when any muscle is converted into meat, the glycogen is hydrolysed by anaerobic glycolysis into lactic acid, manifested by a pH decrease to values between 5-6. Immediately after, bacteria start decomposition of the most straightforward consumable substances. At this step, despite the wide variety of microorganisms and substrates, the spoilage process of all proteinaceous foods is similar, being mainly related to the classes of precursors commonly present in these matrices. Firstly, the degradation of sugars and fats by bacteria produces molecules able to migrate from the meat to the headspace. In this phase, the volatile organic compounds, VOCs, are dominated by EtOH, 3-methyl-1-butanol and free fatty acids, mainly acetic acid. Consequently, the acidity in the headspace remains at a level around 5. Until this phase lasts, protein foods are safe products. A proper sensing device would exhibit the ability to recognise this stage, and, oddly, no attention was deserved to signal it. Only when glucose and its direct metabolites are depleted, the catabolism of proteins starts, producing an assortment of amines, known as the biogenic amines, BAs, sulphur compounds and thiols. As the spoilage proceeds, the discolouration and the production of off-odours make this stage manifested. By now, the by-products are toxic, and meat consumption at this stage could be a severe hazard. A sensing system must be able to recognise this step visibly.

The overwhelming majority of papers on this topic have stubbornly focused on sensing BAs in the headspace. Undoubtedly, BAs are present within the meat. They can be quantified by instrumental methods [27, 28] and titrated indirectly through methods that define the total volatile basic nitrogen, TVBN. The standard method is based on the digestion of meat in perchloric acid or trichloroacetic acid to extract the basic substances, transforming them into ammonium salt. A distillation in strong basic medium allows quantifying all the basic nitrogen as ammonia. [29].

Nevertheless, their existence in the solid does not imply finding them into the headspace, and BAs' fate is less dynamic than expected. This assumption has been proven when the volatile composition is determined. [30–32] If BAs were present in the volatile mixture, the pH would increase at definitively basic values. However, it is not the case. The golden point is: BAs are weak bases, many of them involved in one or more protonation equilibria. During degradation, only neutral molecules can pass into the headspace. In the case of meat and other biological matrices, the pH is buffered at a value around seven. It is evident that, at these pH values, all the amines present a positively charged protonated form and cannot pass into the vapour phase. Also in advanced degradation, the acidity of meat slightly decreases, never exceeding neutrality, preventing amines occurrence in the headspace. When receptors' choice is based on synthetic or enriched samples tests, indicators with colour change at basic pH are often selected [15, 24]· However, this strategy fails when applied to real sample at home conditions (meat kept for a limited number of days in the refrigerator). It must be underlined that the fish represents the only exception: some amines are present also in the headspace due to different nitrogen excretion mechanism. Possibly for this reason, in the literature, many papers have focused on sensing the fish spoilage, [14, 18, 19, 33]. Nevertheless, also for fish, neither the pH reaches basic values, nor BAs different from trimethylamine can be detected in the headspace. The following will explain why the assumption of BAs sensing in volatilome has developed throughout the last twenty years of literature. This myth also persists when the devised solution would refuse the assumption.

2. The creation of a myth

The idea of the possible control of food quality based on sensors started to be discussed in the literature from the late nineties. Yano and coworkers presented in 1996 a biosensor based on electrochemical detection of putrescine and xanthine. [34] Another enzyme sensor array was proposed for the determination of BAs inside the meat. [35] The possibility to apply molecular recognition also for volatile amines detection was explored by [36]. These papers intended to underline the signal's versatility based on a development, or a change, of a colour. As long as real samples analysis is concerned, BAs' detection was mainly dedicated, as already underlined, to fish meat. [18, 19, 33] For instance, Pacquit and collaborators [33] developed a colourimetric sensor based on bromocresol green as a receptor. The sensor is responsive to the headspace composition and changes its colour from yellow to green, as the spoilage proceeds. The authors assessed that BAs' production causes pH change shown by the sensor, but they did not explain why they selected an indicator with $\log K_a$ less than five. They correlated the colour change with the bacterial growth patterns in cod and whiting fish samples, which is indirectly true, but they did not care to attest the presence of amines in the headspace. Electronic tongues for freshness analyses appeared in those years and the contribution of Gills and coworkers, who presented an array of sensors [14] for fish freshness assessment, is an example. This paper allows to point out another issue often encountered in the attempt to propose practical applications. The research is well organised, the array made of sixteen electrodes is well presented, the validation analyses performed. Nevertheless, it does not make sense to have a device able to assess spoilage to start from the 6th- 8th days of sea bream fillet chilled stored. After one-week chilled storage, there is no need for sixteen sensors, not even one, to assess the fish freshness. The point deserves a comment; there is much room between a successful experiment and a possible workable application based on that device, that could operate under real-life conditions. It is also the case where a chameleon probe to assess the biogenic ammines was presented. [37] The successful recognition was not performed in the vapour phase, but in solution with synthetic samples spiked with known BAs amounts. Also, Soga in 2013 [38] proposed an inkjet-printed paper-based colourimetric array to discriminate volatile ammines, but it was tested on vapours of seven primary amines. Solinas and collaborators [16] presented a colourimetric sensors array to check the meat freshness from the real samples' VOCs analyses. However, the final two sensors array could discriminate chicken spoilage after 0-3 days from that after 5-7 and 10-12 days, stored in chilled conditions and modified atmosphere containers (30%CO_2 and 70%N_2). They applied the same device also to pork sausages. [39] The authors demonstrated that the change of colour is correlated with the microbial analysis. They clearly stated that one sensor specifically reacts with biogenic amines and aminated compounds present in the headspace. In [40] the researchers set up a portable optoelectronic nose applied to beef, chicken, pork, fish and shrimps samples, calibrated with H_2S, $(CH_3)_2S$, trimethylamine and cadaverine, but not proving all these analytes being present in the headspace of the investigated samples. The optical sensor design to detect amines during food spoilage, based on the colourimetric recognition of ammonia and biogenic amines, [41] works with all the most common BAs coming from a pH 8.4 buffer solution turning from green to red. The authors claim that the colour change on real meat samples is not equally good because the receptor's logK_a is 6.1, not enough to detect BAs. As will be argued, the reason is that in the headspace there are not BAs at all. They declare to look for a receptor with a lower logK_a. That device will work, not because of interaction with BAs, but because acid development ends, as highlighted below. A sensor designed explicitly for colourimetric detection of thiols and biogenic amines was developed by some of the same authors but tested only in solution on spiked samples. [42] The misleading idea is to deal with biogenic amines that can turn an indicator in its acid form into the basic one. In the paper of Kuswandi, two sensors based on methyl red, logK_a around 4.8, and bromocresol purple logK_a around 6.3, both placed in their acid form [20] are indeed able to sense the spoilage. They proved that the meat's pH, neither after 14 days of storage in a fridge, exceeds 7.30. If this receptor work, it is because it is not sensing amines, as authors claim. The sensor reveals that acid substances are no longer produced. Indeed, even if basic substances were now released, they are immediately neutralised by the buffer systems, preventing further pH increase and BAs volatility. The beef of the example is declared eatable after seven days of storage in the fridge. One week storage seems definitive too long, even if the meat, according to their description, does not come from a maturation process. Such a process is mandatory to reach the desired tenderness, especially with beef meat. As a comment, it is difficult comparing the performance of devices designed to respect different domestic regulatory regimes. As it happens in many fields, the acceptance of common regulation will improve the comparability and the reciprocal improvement in the research.

More recently in many colourimetric sensors, the conviction of sensing volatile amines persists, even if Chen, for instance, developed a device based on a selection of dyes that can assess the change of the pH, in the pork meat. The change the authors claim is correlated to BAs inside the meat since they estimated the TVNB. To be underlined that the pH ranged from 5.8 to 6.7 at the 8th degradation day. After one week in the fridge, it is hard to imagine the meat as eatable [17].

In a more recent paper, the authors synthesised pH indicators specifically for sensing ammonia vapours. They claim that the sensors react to ammonia produced over crabs, and one of them changes its colour between 7 and 12. How it could turn its colour over shrimps remains unclear. [24] A chlorophenol red-based sensor is also successfully employed to sense the spoilage but once more not because of amines in the headspace, as claimed [25].

3. Acid–base equilibria of BAs

A reliable picture of what develops over a piece of meat is acquired when the BAs' acid–base properties are carefully analysed. It suddenly becomes clear why

BAs cannot be found in the headspace over the piece of meat, if not at a miserable amount, at least under the average storage period, as it could happen to meat/fish in a domestic fridge. In **Figure 3**, the acid–base speciation schemes of the most common BAs are shown. They were calculated based on equilibrium constants reported in the literature.

Specifically see, for spermidine and putrescine, [43] cadaverine, [44] histamine, [45] spermine, [46] tyramine, [47] 2 ph-ethyl amine, [48] tryptamine, [49] trimethyl-amine. [50] In **Figure 3**, the vertical dotted lines are drawn in correspondence of the pH values where the fraction of L is equal to 0.1. It is worth to note that for spermidine, cadaverine, spermine, tyramine, the volatile L form reaches 10% only at pH higher than 10, putrescine at pH = 9.80, 2 ph-ethylamine and tryptamine at pH = 9.20. So, even if the catabolism of protein produces BAs into the meat, they cannot be part of VOCs if not in an insignificant amount. Trimethyl-amine, which is typically a by-product of the fish meat, is the only one that has a fraction of L equal to 10% at pH = 8.80, but this fraction represents 1% of the total at pH = 7.4. As underlined elsewhere, the fish metabolism differs from that land vertebrates, having the extraction system mainly through gills and skin. That is why the fresh fish has a fishy odour, and it explains why BA was found in the fresh cod fillets headspace. It is the only case where a headspace device could sense a BA.

The conclusions driven by these calculations, clearly highlighted by the graphs of **Figure 3**, are nothing but corroboration of the experimental evidence reported in papers dealing with real meat samples analyses, both in the solid and in the headspace previously discussed [27, 28, 30–32, 51, 52]. They are presented to make more explicit the real role of BAs in meat spoilage detection, performed by headspace sensors.

Nevertheless, insisting on the idea of looking for BAs in the headspace of spoiled meat can lead to devices that are not correctly centred on the target analytes. The comparison between the by-products dominating the two separated bodies, i.e. the

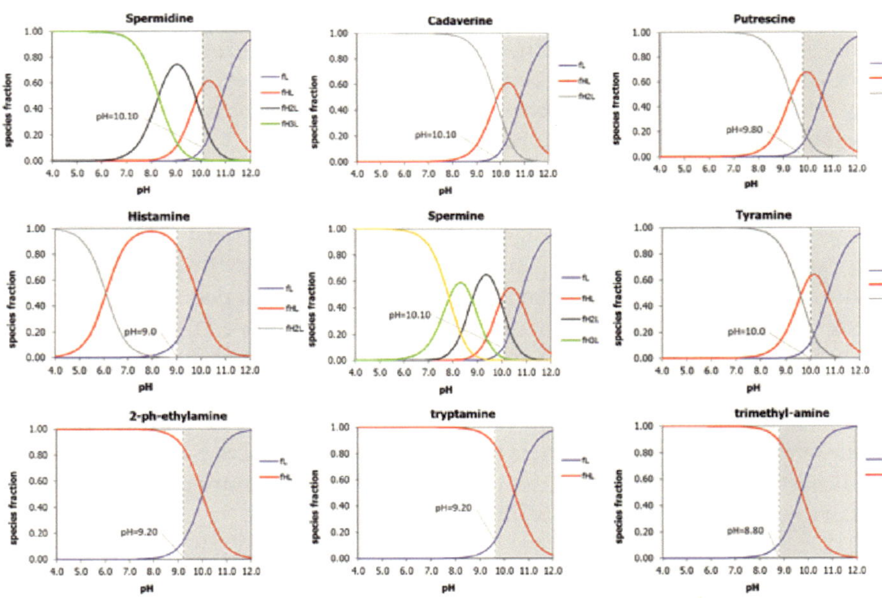

Figure 3.
Acid–base speciation scheme of some BAs calculated from logK_a values found in the literature. [43–50] the vertical line corresponds with the pH values at which the fraction of the fully deprotonated form (L) is 10%. The shadow area represents the L domain, where L fraction is above 0.1.

meat and the VOCs over it, analysed at the same degradation step, offers the best strategy to design a sensor and understand that misleading unambiguously. Here the analysis of the two samples of protein food, such as chicken and cod fillet, at different degradation steps is shown. These were the results of a validation procedure for assessing an array of colourimetric sensors' correct behaviour. The array's performances were explored to develop an intelligent label to signal chicken, beef, pork, and fish spoilage. See reference [4–6]. The device will be described below, and two patents have been deposited to protect the idea. [53, 54] The meat samples were analysed by HPLC-ESI/MS analysis, while the headspace HSSPME/GC–MS analysis. Details of the experimental procedures and the analytical methods can be found in reference [5, 6]. There are many papers where the composition of volatile substances [29–32] and the by-products developed inside the meat [27, 28] are presented. The findings of the experiments reported here are in agreement with this literature. The aim was not to fully characterise the substances produced during the spoilage but to underline the different main classes of substances developed in a meat sample and the atmosphere over it, at the same degradation level.

The results of the analyses are shown in **Table 1** for the chicken meat samples, in **Table 2** for the cod fillets. In the chicken meat case, the qualitative identification of BAs present in the solid was clear enough to distinguish the samples belonging to different degradation steps. In an early stage (analysis done within the first day after purchase), samples were classified as SAFE, and no amines were detected. For samples at the third-day chilled storage, where a slight discolouration and off-odour are perceived, classified as a WARNING zone, 4 out of 7 amines were found. All the BAs under investigation were detected in samples analysed after 4-5 days of storage, classified as HAZARD, when a strong off-odour was also perceived. Opposite, in the case of cod fillets, almost all the BAs can be detected, even in the samples analysed just after purchase. It is well known that fish samples contain amines, even when fresh since the fishy odour is due to volatile amines, in particular trimethylamine.

It must be considered that, due to the high perishability of this food, the production of BAs is very fast and significant even in the time required for the analysis, being the manipulation performed at room temperature and lasting some hours. This drawback originated with the continuous changes in meat composition during analysis was also underlined by Marta Mikš-Krajnik [31].

Nevertheless, a substantial difference can be observed between the samples analysed immediately (SAFE samples) and the ones after three-day storage (HAZARD samples) in the amount of BAs detected. Except for 2-phenylethylamine

BA	Ion(P)	S	Area S	W	Area W	H	Area H
Spermidine	146	—	—	✓	573 723	✓	558 980
Cadaverine	103	—	—	—	—	✓	199 409
Putrescine	89	—	—	—	—	✓	13 577
Histamine	112	—	—	✓	38 082	✓	37 320
Spermine	203	—	—	✓	55 085	✓	45 372
Tyramine	138	—	—	—	—	✓	25 4478
2-ph-ethylamine	122	—	—	✓	32 923	✓	30 311

Adapted with permission from reference [5].

Table 1.
The BAs detected in chicken meat, performed through HPLC-ESI/MS analysis, in correspondence with the three degradation steps, S (safe), W (warning) and H (hazard).

and tryptamine, the pick areas of amines almost double from SAFE samples to the HAZARD ones.

At the same time, the headspace was analysed, too. The results are reported below, in **Tables 3** and **4** for both foods.

In the case of chicken meat, the interpretation is straightforward. Initially, in the samples analysed within the first-day storage (SAFE), very few classes of compounds were detected, mainly acid compounds. For the samples analysed in at the intermediate spoilage, ketones and esters were found, released after the bacterial catabolism of sugars and their derivatives. However, no amines were detected, even if, at this stage, they were present in the meat, see **Table 1**. Nevertheless, more surprisingly, neither in the meat classified as HAZARD, when all the BAs were detected in large amount in the solid, see **Table 1**, amines were found in the headspace, confirming our previous assumption on this topic. [5, 6] Deserve note the presence of indole, in the last samples belonging to the HAZARD stage, originated by bacteria degradation of tryptophan.

In the case of cod fillets, the composition of the headspace in the two samples was very similar, and trimethylamine was present in both samples, nevertheless

BA	Ion(P)	S	Area S	H	Area H
Spermidine	146	✓	13 079	✓	28 800
Cadaverine	103	✓	31 739	✓	62 863
Putrescine	89	—	—	—	—
Histamine	112	✓	8 993	✓	40 463
Spermine	203	✓	12 111	✓	34 185
Tyramine	138	—	—	✓	19 399
2-ph-ethylamine	122	✓	53 750	✓	68 334
Tryptamine	161	✓	52 519	✓	63 562

Adapted with permission from reference [5].

Table 2.
The BAs detected in cod fillets through HPLC-ESI/MS analysis, at two degradation steps, S (safe) and H (hazard).

	S	W	H
Alcohols	✓	✓	✓
Aldehydes	✓	✓	✓
Ethanol	✓	✓	✓
Acids	✓	✓	✓
Ketones	—	✓	✓
Esters	—	✓	✓
Thiols	—	✓	✓
Biogenic amines	—	—	—
Indole	—	—	✓

Adapted with permission from reference [5].

Table 3.
The class of substances detected in the headspace of chicken meat, performed through HSSPME/GC–MS analysis, in correspondence with the three degradation steps S (safe), W (warning) and H (hazard).

	S	H
Alcohols	✓	✓
Aldehydes	✓	✓
Ethanol	✓	✓
Acids	✓	✓
Ketones	✓	✓
Esters	✓	✓
Thiols	✓	✓
Trimethylamine	✓	✓
Light volatile amines	—	✓
Biogenic amines	—	—
Indole	—	✓

Adapted with permission from reference [5].

Table 4.
Class of substances detected in the headspace of cod fillets, performed through HSSPME/GC–MS analysis, at two degradation steps, S (safe) and H (hazard).

crucial details were observed. Volatile amines (here defined as ammonia, mono and dimethyl amines) were present in the HAZARD step, but again BAs were not detected; secondly, indole was revealed in the second part of the degradation process, similarly to what observed in chicken meat samples.

The instrumental analysis performed simultaneously, confirmed that BAs were produced, even in large quantity, in the solid when food is no longer eatable. Conversely, BAs are not detected in the headspace at any step. This distribution is caused by the buffers present in the food that leave BAs in their protonated form, preventing their possibility to pass into the vapour phase. As a result, the pH in the headspace shows only a very slight increase and consequently only the receptor with a very low log K_a turned out to be informative to detect the beginning of the final spoilage stage value, showing a complete and glaring conversion to the basic form. Only for fish samples, some small volatile amines are present but at such a low concentration as not to require a different sensing approach from that one proven to be suitable for chicken meat.

4. Monitoring spoilage from a new perspective

Based on the evidence that BAs cannot be present in the headspace, and the protein catabolism is associated with a small increase of the pH, the idea of controlling the entire spoilage can be faced differently. The two separate steps of spoilage can be identified through indicators that change their colour to sense slight pH modification, placing them into the array in the most convenient form.

A clear signal must help to identify the persistence of the safety zone when protein catabolism is still far away. Depending on the type of food, the beginning of the release of dangerous metabolites could arise in a short time, as it happens for fish, or long enough to allow to detect a transition in between. These behaviours reflect a common situation in cooking: the fish turns quickly from being judged eatable, to an undesirable aspect, a discolouration and a strong unpleasant smell. Opposite we all experienced how other meats take longer time to evolve to a final state and, consequently, to widen our indecision about their fate. At least another sensor, or two, must detect this further step or the transition to the final one, well different from the first.

On this knowledge, a first array was tested. It was a sort of "proof of concept" to assess the feasibility to employ the sensor for naked-eye detection of the spoilage. The indicators were here fixed by ion exchange on a cellulose-based sheet. Details can be found in ref. [5, 6, 53, 54].

A fundamental aspect of the strategy deserves a particular focus. The devices were tested on the same plastic trays usually employed in the supermarket to sell meat, keeping food in the domestic refrigerator for a time lasting just a couple of days after the expiration day, never more than 5-6 days, even less in case of fish. This type of meat as prepared in the supermarket, not in modified atmosphere, is supposed to be consumed within three days after purchase, thus much longer monitoring time lengths sounds useless. As already commented, very commonly, in the literature, the observation times are extended to weeks.

To detect the different stages of meat degradation, the receptors must be selected carefully. In **Table 5**, the most common pH indicators are reported. Receptors from one to four, with higher $\log K_{a1}$, can signal the phase characterised by the acid development. If placed in their basic form over the meat, they change in colour of their protonated form, because of the reaction with the acidic substance typical of the first degradation step. It is worth to note that none of them come back to the basic colour when the proteins catabolism starts, neither bromothymol blue. It is another proof that the pH of the headspace is below seven, as already argued. Conversely, the following dangerous transition can be signalled by an indicator with lower $\log K_{a1}$, put in its acid form. The sensor selected was that containing chlorophenol red indicator. It remains unchanged during the first step, but it turns into its basic colour as soon as acids are not produced anymore.

This strategy is presented in the array of sensors reported in [4–6].

The five indicators of **Table 5**, plus another one able to sense sulphur and thiols, based on Ellman's reagent, was all tested, to select a reasoned final triplet.

An example of the pilot experiment is reported in **Figure 4**. More details can be found in the original papers. It was chosen to operate in conditions like that encountered in real life, to obtain a tailored product, that works under condition that a consumer experienced. Most of the meat and fish are sold in supermarkets prepared by the central slaughterhouse into plastic trays (PP) and covered with low permeability polyethene plastic film. The ratio between the amount of meat and the volume of the headspace, the optimum amount of dye able to combine sensitivity and naked-eye detection and the most reliable and efficient sensors preparation procedure were all taken into account.

A naked eyes evaluation of the evolution of the colours identifies the most informative sensors. The development of a glaring colour among different degradation steps is mandatory for a practical application, and some sensors resulted in practice better than others. Nevertheless, the photos of the sensors colour' changes

	$\log K_{a1}$	$\log K_{a2}$
m-Cresol purple	8.32	1.57
o-Cresol red	8.20	1.11
Thymol blue	8.9	1.5
Bromothymol blue	7.1	
Chlorophenol red	6.0	

Table 5.
The selection of golden pH indicators for spoilage meat sensing based on their protonation constants, as found in reference [55, 56].

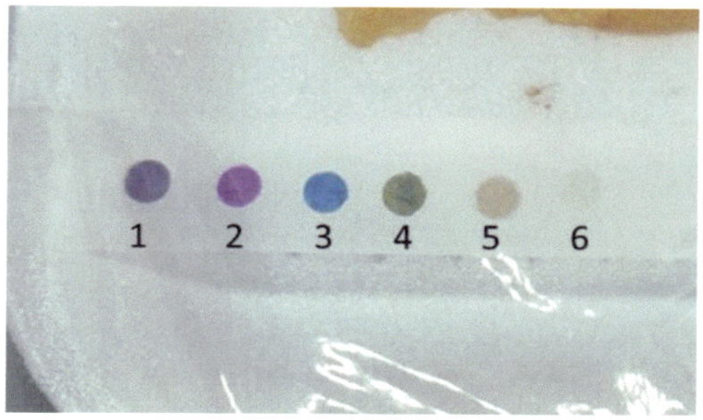

Figure 4.
An example of the array placed over the tray containing the poultry meat, with sensors from one to six based on m-cresol purple (1), o-cresol red (2), bromothymol blue (3), thymol blue (4), chlorophenol red (5), and the Ellman's reagent (6).

were acquired, transformed into RGB triplets, for each sensor at any registered time. The PCA, Principal Component Analysis, well known chemometric tool, was selected to represent the data. Indeed, PCA allows summarising in a few graphs all the information contained in a wide dataset. It is the case of the RGB triplets data collected during the entire degradation. After performing PCA, the most important variables, i.e. the most significant sensors that describe the spoilage, are identified in the so-called loadings plot. In the scores plot, samples with similar properties result close to each other, i.e. the samples at the same degradation step are grouped. Details about PCA could be found elsewhere, but the PCA applied to the present dataset, confirms the qualitative findings. In the case of chicken breast, beef and pork meat, a stage between the safe step and the hazard one can be detected, not for fish. [6] In **Figure 5**, as an example, the PCA models on the first two components for chilled stored chicken samples are shown. Only the three most significative sensors are included. In the loadings plot, the different contribution of the three sensors arises. In the scores plot, the sensors' evolution, from colours describing safe samples placed on the left, toward hazard condition to the right emerges clearly. The model was validated; it is not an artefact. Indeed, external samples are correctly projected into the model, see the blue samples in **Figure 5**. Chemical analyses of samples assigned to the three steps effectively belong to different spoilage stage.

The results confirm that the sensors' change of colours follows the real evolution of volatilome. The array allows a naked eye evaluation, feasible also for a consumer. It means that it is possible to assign a precise colour to each step: these reference colours can be printed on a label. The array inserted into the package containing the meat samples will change its colour according to the spoilage evolution. Comparing the colour of the array with the reference, anyone can decide at home the fate of the meat in the fridge, without any doubt.

As a further evolution of the array, the solid support was changed. [57] In the first cellulose-based array, the dyes were fixed by ion exchange. This easy and cheap construction was chosen for its versatility, low cost, and the possibility to obtain a large number of testing materials was a necessary proof of concept.

The derivatisation on EVOH copolymer, intensely employed by the food packaging industry as an oxygen barrier film, was the choice to covalently bind the receptor on the support. [53, 54] These derivatives are stable, still very cheap and prevent any leaching of the indicator. The final product, obtained in grains, was filmed by a mechanic press under controlled temperature and pressure. From the foils, sensors of 0.5 cm of diameter were obtained using an office puncher. [57] An example of the materials is presented in **Figure 6**. The reference [58] reports the synthesis and the characterisation of the material.

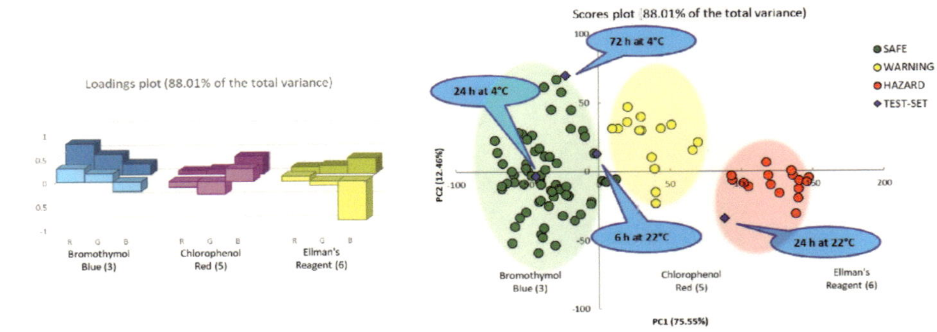

Figure 5.
PCA model on the data set of the RGB data of three indicators, based on the first two components that explain 88.1% of the total variance. On the left, the loadings plot, in foreground values on the PCA1, in the background those on PCA2. On the right, the scores plot. The blue bubbles identified the projection of external samples. The green, yellow, and red shadow areas collect most of the samples defined as safe, warning, and hazard. Adapted with permission from reference [5].

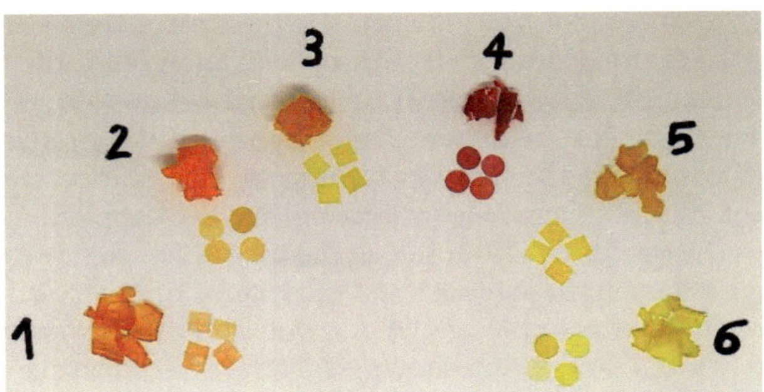

Figure 6.
Grains and miniaturised sensors obtained after functionalisation of EVOH with different sulphonftaleins, from left to right m-cresol purple (1), o-cresol red (2), bromothymol blue (3), thymol blue (4), chlorophenol red (5) and bromophenol blue (6).

These sensors exhibit glaring, highly reproducible colours. As seen in **Figure 6**, the materials are suitable to be tested as meat spoilage sensors. The receptors' encapsulation into the polymeric mainframe, makes indicators fixed into the mainframe weaker acid and their protonation constants moved to higher values, at least one order of magnitude. [57, 58]. This evidence was taken into account and the array modified. The acids developed over the tray can still be revealed by any receptor with logK_{a1} higher than seven, see **Table 5**. The first four indicators, placed in their basic form, are still, or even more, suitable candidates. Indeed, they very quickly react with the acidic volatilome, typical of the first save degradation step. The chlorophenol red-base sensor, previously employed to detect the transition toward spoiled meat does not work anymore. In the EVOH derivative, it changes the colour at too high pH. Solution tests demonstrated it very clearly, as shown in **Figure 7** At pH equal to seven, the dye in solution is definitely in its basic colour, while when inserted in the solid, equilibrated at the same pH, it is only in the transition from its acidic to its basic colour.

The solution that accounts for the sensors' acid–base properties changes is to develop a dual-sensor device.

The acid–base equilibria of *o*-cresol and the colours associated with the three different species are reported in the upper part of **Figure 8**. The advantage is that in the EVOH derivative, being the protonation constant higher, the fully protonated form is obtained in less extreme pH values and once obtained is stable at the air.

The idea of the dual-sensor based on EVOH derivatised with *o*-cresol is conceptualised in the lower part of **Figure 8**.

Suppose the *o*-cresol red-base sensor is placed in its fully protonated and deprotonated forms, as reported in the last picture on the right of **Figure 8**. Once the sensor is filmed, the spot at the two extreme acidic and basic forms exhibits almost the same violet colour. On the contrary, the yellow colour of the mono-protonated form almost disappears. The double *o*-cresol based sensor works as an on/off device. The first transition is signalled by the basic form of the indicator, which turns off. It means that the fully deprotonated form reacts with acid becoming the pale yellow of the monoprotonated form. As long as acids developed in the headspace, during the first phase, the other spot placed in fully protonated form, remains violet; in the on/off terms, switched on. The final stage is signalled by the last spot shutdown, due to the slight increase of the pH environment.

The first tests on real chicken samples seem very promising. This device's advantages will be evident when moving to the lab product to a market validation prototype. There is not any need to carefully print the reference colours and for the consumer to compare three different colours. The intelligent label can be read as binary on/off signal. Other protein foods must be tested, and further experiments have been already untaken. Presently, some features, such as low cost, easy implementation, absence of leaching, have been achieved.

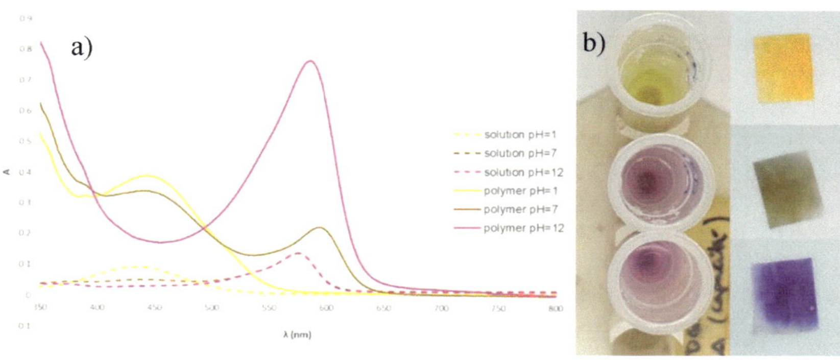

Figure 7.
UV–vis spectra (a) and corresponding photographs (b) of chlorophenol red-solution 11 μM (dashed lines) and the corresponding colour of the chlorophenol red-EVOH@ (solid lines) after equilibration at pH 1 (up), 7 (Centre) and 12 (down).

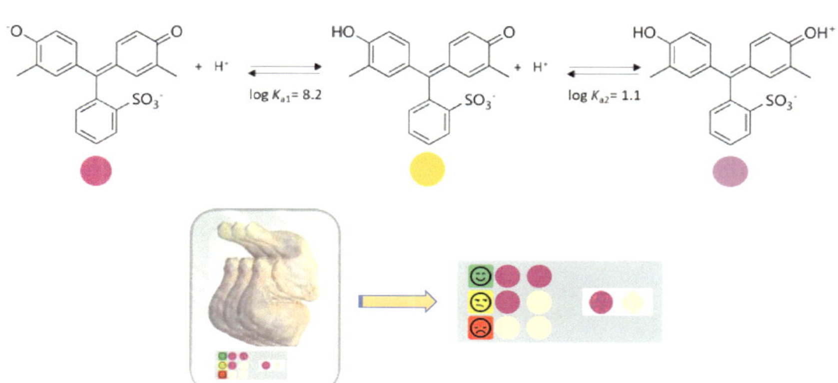

Figure 8.
In the upper part, the different acid–base forms of the o-cresol red and their colours. In the lower part, the new prototype of the intelligent label inserted in the tray with its magnification.

The device illustrated here is at an advanced level of development compared to most of the ideas proposed in the last twenty years' literature, aimed at developing smart labels to be inserted in food packaging. In the case studies presented, the authors expect the idea to develop into a concrete product. The mission is to answer two main issues behind the continuous development and increase of the proteinaceous foods market. Food poisoning is still a public health concern: such labels could substantially reduce the consumption of failed food, especially when it origins at home, after purchase. A not less critical point, clear indication of the spoilage level could limit food waste and allow food consumption beyond the expiry date. Indeed, when the food is stored in optimal conditions, the expiry date induces to dismiss food still edible since the date is established conservatively. In our economy, the waste of still eatable food represents a substantial not admissible social cost.

5. Conclusion

The misleading idea to find BAs in the headspace was born and persist in the literature that deals with the colourimetric sensors designed to follow the protein food spoilage. Often the researchers assess to detect biogenic amines. However, they selected receptors with logK_a values low enough to sense the slight increase in pH, not caused by the presence of BAs in the volatilome. Indeed, even in an advanced decomposition state, the meat pH values never exceed neutrality, preventing BAs from being present in their volatile form.

There is a great demand to have optodes for degradation monitoring. The fundamental issues in the design of these sensors are summarised below.

First start point should be an accurate description of the problem, which, on the contrary, has often been rough. In this field, maybe better results in less time would have reached. Nowadays, many optodes, present in the literature, are close to the objective, but often unknowingly, as underlined above.

The second point that deserves attention is that candidates for intelligent labels must be developed under conditions close to that a consumer faced real life. The research must go on with proof of concept, but dealing with applied science, greater attention to boundary conditions is desirable.

The device here presented fulfils all the issues and is an advanced development phase, even if preindustrial. For instance, EFSA authorisation to declare the labels suitable for food packaging is pending; the pilot production in high quantity will start only within this year to deeply evaluate the industrial scalability. The research is presently focused on compostable or eatable materials, to be used as solid support. The aim is to offer a green alternative to plastic based EVOH derivatives. The environmental impact must become a pressing concern when new products move from the lab into the market, being the footprint into the world always as light as possible.

There is common awareness about the social and economic impact of the intelligent systems in the market. The authors hope to see reliable smart devices into the market that give food quality granted and have made a contribution in this sense.

Acknowledgements

The authors thank the MIUR of Italy for funding Lisa Rita Magnaghi's PhD grant, and PRIN2015 "Multiple equilibria in natural and biological fluids: from speciation to selective sequestering" (code 2015MP34H3_004) for valuable financial support.

Conflict of interest

The authors declare no conflict of interest.

Notes/thanks/other declarations

Authors express their gratitude to Federica Capone, PhD Student, for last development's contribution in this research and help in editing. If not indicated, figures are produced originally for this chapter and photos are of the authors' database.

Author details

Raffaela Biesuz* and Lisa Rita Magnaghi
Department of Chemistry, University of Pavia, Pavia, Italy

*Address all correspondence to: rbiesuz@unipv.it

References

[1] Halonen N, Pálvölgyi P S, Bassani A, Fiorentini C, Nair R, Spigno G, Kordas K. Bio-Based Smart Materials for Food Packaging and Sensors – A Review. Front. Mater. 2020; 7,82. DOI:10.3389/fmats.2020.00082.

[2] Ahmed I, Lin H, Zou L, et al. An overview of smart packaging technologies for monitoring safety and quality of meat and meat products. Packag Technol Sci. 2018; 31:449-471. DOI: 10.1002/pts.2380.

[3] Wang S, Liu X, Yang M, Zhang Y, Xiang K, Tang R. Review of Time Temperature Indicators as Quality Monitors in Food Packaging. Packag. Technol. Sci. 2015; 28: 839-867. DOI: 10.1002/pts.2148.

[4] Magnaghi L R, Alberti G, Quadrelli P, Biesuz R. Development of a Dye-Based Device to Assess Poultry Meat Spoilage. Part I: Building and Testing the Sensitive Array. J. Agric. Food Chem. 2020; 68,45: 12702-12709. DOI: 10.1021/acs.jafc.0c03768.

[5] Magnaghi L R, Alberti G, Capone F, Zanoni C, Mannucci B, Quadrelli P, Biesuz R. Development of a Dye-Based Device to Assess the Poultry Meat Spoilage. Part II: Array on Act. J. Agric. Food Chem. 2020; 68,45: 12710-12718. DOI: 10.1021/acs.jafc.0c03771.

[6] Magnaghi L R, Capone F, Zanoni C, Alberti G, Quadrelli P, Biesuz R. Colorimetric Sensor Array for Monitoring, Modelling and Comparing Spoilage Processes of Different Meat and Fish Foods. Foods. 2020; 9: 684; DOI: 10.3390/foods9050684

[7] Hannah Ritchie (2017) - "Meat and Dairy Production". Published online at OurWorldInData.org. Retrieved from: 'https://ourworldindata.org/meat-production' [Online Resource].

[8] Havelaar AH, Kirk MD, Torgerson PR, GibbHJ, Hald T, Lake RJ, et al. World Health Organization Global Estimates and Regional Comparisons of the Burden of Foodborne Disease in 2010. PLoS Med. 2015; 12(12): e1001923; DOI:10.1371/journal.pmed.1001923

[9] FAO reports Available from http://www.fao.org/3/a-i3991e.pdf- pg 67

[10] Mills A, Hawthorne D, Grahm A, Lawrie. Novel time-temperature and "consume-within" indicator based on gas diffusion. Chemm Comm. 2016; 52:13987-1399. DOI:10.1039/c6cc07906g.

[11] Yusufu D, Mills A. A colourimetric vacuum air-pressure indicator. Analyst. 2019; 144: 5947. DOI: 10.1039/c9an01507h.

[12] Wojonowski W, Majchrzak T, Dymerski T, Gebicki J, Namiesnik J. Electronic noses: powerful tools in meat quality assessment. Meat Science. 2017; 131: 119-131. DOI: 10.1016/j.meatsci.2017.04.240.

[13] Cetó X, Voelcker N H, Prieto-Simon B. Bioeletronic tounges: new trends and application in water and food analysis. Biosensors and bioelectronics. 2016; 79:608-626. DOI:10.106/j.bios.2015.12.075

[14] Gil L, Barat J B, Escriche I, Garcia-Breijo E, Martínez-Máñez R, Soto. An electronic tongue for fish freshness analysis using a thick-film array of electrodes. Microchim Acta. 2008; 163: 121-129. DOI: 10.1007/s00604-007-0934-5.

[15] Chen Q, Hui Z, Zhao J, Ouyang Q. Evaluation of chicken freshness using a low-cost colorimetric sensor array with AdaBoost–OLDA classification algorithm. LWT - Food Science and

Technology, 2014; 57: 502-507 DOI: 10.1016/j.lwt.2014.02.031.

[16] Salinas Y, Ros-Lis J V, Vivancos J L, Martinez-Manes R, Marcos M D, Aucejo S, Herranz N, Lorente I. A novel colometric sensor array for monitoring fresh pork sausages. Food Control. 2014; 35: 166-176. doi: 10.1016/j.foodcont.2013.06.043.

[17] Chen H, Zhang, M, Bhandari, B, Yang, C. Development of a novel colorimetric food package label for monitoring lean pork freshness. LWT - Food Science and Technology. 2019; 99: 43-49. DOI 10.1016/j.lwt.2018.09.048.

[18] K. Crowley, A. Pacquit, J. Hayes, King Tong Lau and D. Diamond, "A gas-phase colorimetric sensor for the detection of amine spoilage products in packaged fish," SENSORS, 2005 IEEE, Irvine, CA, 2005, pp. 4 pp.-, doi: 10.1109/ICSENS.2005.1597809.

[19] Pacquit A, Frisby J, Diamond D, Tong Lau K a, Alan Farrell A, Quilty B, Diamond D. Development of a smart packaging for the monitoring of fish spoilage. Food Chemistry. 2007; 102: 466-470. DOI: 10.1016/j.foodchem.2006.05.052

[20] Kuswandi B, Nurfawaidi A. On-package dual sensors label based on pH indicators for real-time monitoring of beef freshness. Food Control. 2017; 82: 91-100. DOI: 10.1016/j.foodcont.2017.06.028.

[21] Dudnyk I et al. Edible sensors for meat and seafood freshness. Sensors and Actuators B. 2018; 259:1108-1112. DOI: 10.1016/j.snb.2017.12.057.

[22] Ezati P et al. Intelligent pH-sensitive indicator based on starch-cellulose and alizarin dye to track freshness of rainbow trout fillet. International Journal of Biological Macromolecules. 2019; 132: 157-165. DOI: 10.1016/j.ijbiomac.2019.03.173.

[23] Ding L, et al. A naked-eye detection polyvinyl alcohol/cellulose-based pH sensor for T intelligent packaging. Carbohydrate Polymers. 2020; 233: 115-859. DOI: 10.1016/j.carbpol.2020.115859.

[24] Zhang H, Hou A, Xie K, Gao A. Smart color-changing paper packaging sensors with pH sensitive chromophores based on azo-anthraquinone reactive dyes. Sensors and Actuators B: Chemical,2019; 286: 362-369 DOI:, 10.1016/j.snb.2019.01.165

[25] Liu B, Gurr P A, QiaoG G Irreversible Spoilage Sensors for Protein-Based Food. ACS Sensors 2020; 5: 2903-2908 DOI: 10.1021/acssensors.0c01211

[26] O'Halloran G R, Troy D J, Buckley D J. The relationship between early post-mortem pH and the tenderisation of beef muscles. Meat science. 1997; 45,2: 239:251. DOI: 10.1016/S0309-1740(96)00074-5.

[27] Sirocchi V, Caprioli G, Ricciutelli M, Vittori S, Sagratini G. Simultaneous determination of ten biogenic amines in meat by liquid chromatography-tandem mass spectrometry (HPLC-MS/MS). J. Mass Spectrom. 2014; 49: 819-825. DOI: 10.1002/jms.3418.

[28] Vinci G, Antonelli M L. Biogenic amines: quality index of freshness in red and white meat. Food Control. 2002; 13: 519-524. DOI: 10.1016/S0956-7135(02)00031-2.

[29] Antonacopoulos, N., Vyncke, W. Determination of volatile basic nitrogen in fish: A third collaborative study by the West European Fish Technologists' Association (WEFTA). Z Lebensm Unters Forch 1989;189: 309-316. DOI:10.1007/BF01683206

[30] Mikš-Krajnik M, Yoon Y J, Yuk H G. Detection of volatile organic compounds as markers of chicken breast spoilage using HS-SPME.GC/MS-FASST. Food

Sci. Biotechnol. 2015; 24: 361-372. DOI: 10.1007/s10068-015-0048-5.

[31] Mikš-Krajnik M, Yoon Y J, Ukuku D O, Yuk H G. Identification and Quantification of Volatile Chemical Spoilage Indexes Associated with Bacterial Growth Dynamics in Aerobically Stored Chicken. J. Food Sci. 2016; 81: 2006-2014, DOI: 10.1111/1750-3841.13371.

[32] Casaburi et al. Bacterial populations and the volatilome associated to meat spoilage. Food Microbiology. 2015; 45: 83-102. DOI: 10.1016/j.fm.2014.02.002.

[33] Pacquit A, Tong Lau K, McLaughlin H, Frisby J, Quilty B, Diamond D. Development of a volatile amine sensor for the monitoring of fish spoilage. Talanta. 2006; 69: 515-520 DOI:10.1016/j.talanta.2005.10.046

[34] Yano Y, Yokoyama K, Tamiya E, Karube I. Direct evaluation of meat spoilage and the progress of aging using biosensors. Analytica Chimica Acta 1996; 320: 269-276DOI: 10.1016/0003-2670(95)00543-9

[35] Lange J, Wittmann C. Enzyme sensor array for the determination of biogenic amines in food sample. Anal. Bionalal. Chem. 2002; 327: 276-283. DOI: 10.1107/s00216-001-1130-9

[36] Suslick K S et al. Colorimetric sensor arrays for molecular recognition. Tetrahedron. 2004; 60: 11133-11138. DOI: 10.1016/j.tet.2004.09.007

[37] Steiner M S, Meier R J, Duerkop A, Wolfbeis O S. Chromogenic sensing of biogenic amines using a chameleon probe and the Red-Green_Blue readout of digital camers images. Analytical Chemistry. 2010; 8402-8405. DOI: 10.1021/ac102029j.

[38] Soga T, Jimbo Y, Suzuki K, Citterio D. Inkjet-printed paper-based colorimetric sensor array for the discrimination of volatile primary amines. .Analytical Chemistry. 2013.85:8973-8978. DOI:10.1021/ac402070z

[39] Salinas Y, Ros-Lis J V, Vivancos Y-L, Martínez-Máñez R, Marcos M D, Aucejo S, Herranz N, Lorente I, Garcia E. A novel colorimetric sensor array for monitoring fresh pork sausages spoilage. Food Control 2014; 35: 166-176. DOI:10.1016/j.foodcont.2013.06.043

[40] Li Z, Suslick K S. Portable optoelectronic nose for monitoring meat freshness. ACS Sens. 2016; 1: 1330–1335. DOI: 10.1021/acssensors.6b00492.

[41] Schaude C, Meindl C, Fröhlich E, Attard J, Mohr G J. Developing a sensor layer for the optical detection of amines during food spoilage. Talanta. 2017; 170: 481-487. DOI: 10.1016/j.talanta.2017.04.029.

[42] Mastnak T, Lobnik A, Mohr GJ, Turel M. Design and Characterization of Dicyanovinyl Reactive Dyes for the Colorimetric Detection of Thiols and Biogenic Amines. Sensors. 2018; 18: 814. DOI: 10.3390/s18030814.

[43] De Robertis A, Foti C, Sammartano S. Dependence on Ionic Strength of Polyamine Protonation in NaCl Aqueous Solution. J. Chem. Eng. 2001; 46:1425 DOI: 10.1021/je010143g.

[44] Cascio S, De Robertis A, Foti C. Protonation of Diamines H2N–(CH2) n–NH2 (n = 2–10) in NaCl Aqueous Solution at Different Ionic Strengths. J. Chem. Eng. Data 1999; 44: 735-738 DOI:10.1021/je980311d

[45] Mohan M S, Bancroft D, Abbott E H Thermodynamic study of the formation of some mixed-ligand complexes of copper (II). Inorganic Chemistry 1979; 18: 344-346 DOI:10.1021/ic50192a027

[46] De Stefano C, Foti C, Sammartano S. Speciation of low

molecular weight ligands in natural fluids: protonation constants and association of open chain polyamines with the major components of seawater. Anal.Chim.Acta 2000; 418:43. DOI: 10.1016/S0003-2670(00)00945-4.

[47] Kiss T, Tóth B. Microscopic dissociation processes of some tyrosine derivatives. Talanta 1982; 29, 539-544, DOI:10.1016/0039-9140(82) 80213-0.

[48] Bunting J, Stefanidis D. A systematic entropy relationship for the general-base catalysis of the deprotonation of a carbon acid. A quantitative probe of transition-state solvation. J.Am. Chem.Soc. 1990; 112:779. DOI: 10.1021/ja00158a043.

[49] Weber O A, Simeon V. Tryptamine, 5-hydroxytryptamine and 5-hydroxytryptophan complexes of proton and some divalent metal ions. Journal of Inorganic and Nuclear Chemistry. 1971; 33: 2097-2101

[50] Takahashi A, Natsume T, Koshino N. Speciation of trimethyltin(IV): hydrolysis, complexation equilibria, and structures of trimethyltin(IV) ion in aqueous solution. Can. J. Chem. 1997; 75:1084. DOI: 10.1139/v97-128

[51] Nychas G-J E, Tassou, C C. Spoilage process and proteolysis in chicken as detected by HPLC. J. Sci. FoodAgric. 1997, 74, 199-208. DOI: 10.1002/(SICI)1097-0010(199706)74:2<199::AID-JSFA790>3.0.CO;2-4

[52] Nychas G-J E, Skadamis P N, Tassou, C C Koutsoumanis K P. Meat spoilage during distribution. Meat Science 2008; 78: 77-89. DOI:10.1016/j.meatsci.2007.06.020

[53] Biesuz R, Quadrelli P, Magnaghi L R. Sensori per la Valutazione della Qualità di Prodotti Alimentari a Base di Carne. Patent 102019000000464. March 19, 2019.

[54] Biesuz R, Quadrelli P, Magnaghi L R. Sensors for the Evaluation of the Quality of Meat-Based Food. WIPO PCT/IB2020/052998. March 30, 2020.

[55] Casula R, Crisponi G, Cristiani F, Nurchi VM, Casu M, Lai A. Characterization of the ionisation and spectral properties of sulfonephtalein indicators. Correlation with substituents effects and structural features. Talanta. 1995; 40: 1781-1788. DOI: 10.1016/0039-9140(95)01559-T.

[56] Aragoni MC, Arca M, Crisponi G, Nurchi VM, Silvagni R. Characterization of the ionisation and spectral properties of sulfonephtalein indicators. Correlation with substituent effects and structural features. Part II. Talanta. 1995; 42: 1157-1163. DOI: 10.1016/0039-9140(95)01559-T.

[57] Capone F. EVOH based sensors for monitoring protein food spoilage [thesis]. Università degli Studi di Pavia; 2020.

[58] Magnaghi L R, Alberti G, Milanese C, Quadrelli P, Biesuz R. Naked-Eye Food Freshness Detection: Innovative Polymeric Optode for High-Protein Food Spoilage Monitoring, accepted for publication on Food Sci& Tech. DOI: 10.1021/acsfoodscitech.0c00089

Chapter 7

Cis/Trans-Fatty Acid Content of Red Meats and the Related Effects on Meat Quality and Human Health

Edward C. Webb

Abstract

Red meats are often criticized as unhealthy based on their perceived high-fat content and saturated fatty acid composition. Uncertainties about the fatty acid composition and trans-fatty acid contents may discourage consumers to eat red meat, especially those living with non-communicable diseases such as cardiovascular diseases, hypertension and obesity (e.g. the metabolic syndrome). Previous studies have investigated the factors that influence the fat content and fatty acid composition of red meats, including the effects of species, age, nutrition, sex, production systems and growth promotants in animals, but the trans-fatty acid content of red meat has not been well studied to date. The purpose of this chapter is to review the fat content and fatty acid composition of red meats, with specific reference to its cis/trans-fatty acid content. Representative samples of beef sirloin steaks (n = 60) and lamb loin chops (n = 80) (the lumbar part of the longissimus dorsi muscle) were collected from carcasses from several randomly selected abattoirs in the Gauteng region of South Africa for proximate and fatty acid analyses. Results from this study confirm that the intramuscular fat content of red meats is low compared to most fat-containing processed foods. The lean component of beef and lamb contain a trivial proportion of TFA's, consisting of vaccenic acid, rumenic acid and conjugated linoleic acid (CLA) isomers. The CLA's in red meat are beneficial due to their antioxidant and anti-carcinogenic properties, so they should not strictly be considered in the TFA definition. This means that the TFA's in red meats are negligible and pose no harm to human health. Labelling of red meats should be improved to convey this information to consumers.

Keywords: Red meat, animal fat, saturated fatty acids, unsaturated fatty acids monounsaturated fatty acids, polyunsaturated fatty acids, cis-, trans-fatty acids, conjugated linoleic acid, CLA

1. Introduction

Red meats are tarnished as unhealthy due to their perceived high-fat content and saturated fatty acid composition [1]. Uncertainties about the trans-fatty acid content of red meats may deter consumers to eat red meat, especially those living with non-communicable diseases such as obesity, hypertension and cardiovascular

IntechOpen

diseases. Much of the criticism against fats in red meats is unfounded and usually poorly understood. Some of these paradoxical issues and perceptions about animal fats and their related effects on meat quality, human health and consumer perceptions have been reviewed in detail previously [1]. However, there is little scientific information about the trans-fatty acids in beef and lamb or the factors that affect its content.

Trans-fatty acids (TFA) are the sum of all unsaturated fatty acids that contain one or more isolated, non-conjugated double bond in the trans geometric configuration [2], (excluding CLA's, which are conjugated). According to the European Food Safety Authority (EFSA), TFA's do not serve any vital functions in the human body because they are neither synthesized nor required by the human body [3]. Health professionals recommend a reduction in overall consumption of saturated fatty acids (SFA), TFA and cholesterol, while increasing intake of n-3 polyunsaturated fatty acids (PUFAs) [4, 5].

The harmful health effects related to TFA intake are specifically those associated with Coronary Heart Disease (CHD), Cardiovascular Disease (CVD) and related diseases [3, 6, 7]. Intake of TFA that exceeds 5 grams per portion (100 g) is associated with an increased risk of CHD [8]. The TFA's in the human diet generally originate from either the partial hydrogenation of vegetable oils and to some extent fish oils, such as deep-fat frying cooking techniques and other heat treatments used in the fast-food industry, or from a small number of natural types of TFA's which are synthesized from polyunsaturated fatty acids (PUFA) by anaerobic bacteria in the rumen of cloven-hoofed animals [9–11].

The TFA content of some industrial foods is reportedly as high as 60% of total fatty acid content and sometimes even higher [11]. In comparison, the ruminant derived TFA's found in ruminant fat, seldom exceed 6%, e.g. the TFA content of milk fat range from 4–6% [11]. In ruminant fat, up to 20% of the TFA content may consist of the C16:1 trans-isomer range, which is not found in industrial TFA profiles [12]. The most common TFA-isomers are elaidic acid (C18:1), vaccenic acid (C18:1 t-11) and rumenic acid (C18:2) [11, 13].

Vaccenic acid and rumenic acid are the trans isomers that are specific to ruminant derived fats [2]. Research indicates that ruminant derived trans vaccenic acid is not associated with an increased risk of CHD or CVD, because it is readily converted to CLA isomers which have numerous health benefits [14, 15]. Trans unsaturated fatty acids when consumed in excessively large amounts cause elevated plasma LDL cholesterol and a reduced HDL cholesterol state, which is associated with CVD, CHD and Type II diabetes mellitus [16, 17].

Little information is available about the trans-fatty acid content of red meats. The purpose of this chapter is to provide an overview of the composition of fats in red meats (beef, lamb, goat (chevon)) in Section 2, and to present new results about the trans-fatty acid composition of red meats in Section 3 in this chapter.

2. Review of the lipid composition of animal fats

The fats in animal tissues are a subgroup of lipids, which consist mostly of triacylglycerols. Animal fats are stored in different adipose tissue (fat) depots, at different physiological stages of animal growth and development. The predominant adipose tissue depots in animals include the subcutaneous fat (underneath the skin), intramuscular fat and a variety of internal fat depots such as perirenal, cardiac, omental and pelvic fat depots.

The growth of animals is characterized by an allometric accumulation of bone, muscle and fat, in that order, with fat accumulating and maturing slower and later than the other tissues. Different adipose depots mature at different rates, with the intramuscular and subcutaneous fat depots maturing slower and later compared to the internal tissues depots.

Red meat has erroneously being labelled as high in fat, especially saturated fatty acids (SFA) which can be associated with many non-communicable human diseases like the metabolic syndrome, but frankly, lean meat only contains 2–3% intramuscular fat in lean beef [18–20], 5–7% in conventional beef [21–23] and as high as 15% in marbled beef. The intramuscular fat content of lamb varies between 8 to 14% [24, 25]. The predominant saturated fatty acids (SFA) in red meats are C14:0 (myristic acid), C16:0 (palmitic acid) and C18:0 (stearic acid). The intramuscular fat contents of red meats are respectively 80% and 40% lower in conventional and marbled beef, compared to the average fat content of processed foods [26] in the USA. The importance of animal fats is highly underestimated, especially in terms of its source of essential fatty acids, serving as a carrier of fat-soluble vitamins A, D, E and K, synthesis of steroidal hormones, role as metabolic energy source, and contribution to sensory properties such as flavor, aroma and texture in meat [1, 18].

2.1 The nutritional value of fat in meat

The lean component of red meat is a highly digestible and high biological value protein source in the human diet, e.g. the digestibility value of red meat is appreciably higher (94%), compared to whole wheat (86%) or beans (78%) [27]. Red meat has a relatively low fat and sodium content, but with high contents of antioxidants and several other nutritionally important bioactive substances such as taurine, carnitine, carnosine, ubiquinone, glutathione and creatine [27], which are physiologically important because of their antioxidant and anti-inflammatory effects, and immunological, neurological, muscular and retinal functions [28]. Taurine, carnitine, carnosine and creatine are abundant in beef, but do not occur in plant-sourced foods [28].

The nutritional value of fats in meat depends essentially on the ratio between saturated (SFA), monounsaturated (MUFA) and polyunsaturated fatty acids (PUFA), as well as the balance between the n-6 and n-3 fatty acids [29]. Conventional wisdom suggests that the fat composition of red meats is all saturated. Several research groups in Australia [30], Europe [18–20], the United Kingdom [21, 31], and in South Africa [1, 23–25] have studied animal fats extensively. All concur that animals fats in red meat species are certainly not all saturated. This represents perhaps one of the most incorrect perceptions about red meats.

A summary of the composition of SFA, MUFA and PUFA previously reported for meat from cattle, sheep and goats are presented in **Table 1**. These results demonstrate that neither beef, lamb nor goat meat are all saturated. Normal saturation levels of these red meats generally vary between about 45 to 54%, with significant proportions of the mono-unsaturated fatty acid oleic acid, ranging from ca. 38 to 44%. Overall the SFA content of beef is less than 3 g per 100 gram edible portion of raw red meat.

Previous recommendations by medical professionals were to reduce the total fat intake to decrease obesity and lower the risk of coronary heart disease (CHD). However, the recommendations have shifted to the fatty acid composition instead of quantity [1]. For example, it is stressed that the MUFAs and PUFAs are more important than total fat ingestion to reduce the risk of cardiovascular heart disease in middle-aged men [34].

Much of the internal carcass fat is removed during the slaughter process, but subcutaneous and intramuscular fat depots remain part of the meat sold to consumers. The removal of carcass fat is regarded as a wasteful process, but its

Meat species	Fatty acid fraction (w/w%)			Reference
	SFA	MUFA	PUFA	
Beef				
Feedlot steers#	49.6 ± 4.4	38.5 ± 3.4	8.7 ± 1.5	[23]
*Belgian white blue**	46.5 ± 4.2	38.4 ± 4.1	15.0 ± 3.9	[19]
Sheep				
Dorper sheep	52.8 ± 1.9	43.9 ± 1.3	3.3 ± 0.5	[32]
Damara sheep	51.8 ± 1.8	44.3 ± 1.3	3.9 ± 0.2	[32]
Dorper sheep	50.8 ± 2.9	42.7 ± 3.7	4.7 ± 0.9	[24, 25]
Merino sheep	52.2 ± 3.9	40.0 ± 3.2	5.21 ± 1.1	[24, 25]
Dorper (Karoo grass)	52.9 ± 4.5	43.1 ± 3.6	3.6 ± 0.9	[33]
Dorper (Karoo browse)	54.7 ± 2.9	41.3 ± 1.9	3.7 ± 1.5	[33]
Goats				
Boer goats	54.7 ± 2.2	41.9 ± 0.9	3.4 ± 0.4	[32]
Indigenous goats	53.6 ± 2.8	42.5 ± 1.1	3.9 ± 0.4	[32]

SFA-Saturated fatty acids; MUFA-Monounsaturated fatty acids; PUFA-Polyunsaturated fatty acids.
#Conventionally raised beef cattle.
**High lean beef breed.*

Table 1.
Summary of saturated and unsaturated fatty acids previously reported for meat from cattle, sheep and goats.

accumulation in the carcass is an inevitable consequence of finishing livestock to the required body condition (grade) to yield the most acceptable and flavorsome meat.

Several methods and biotechnologies have been employed to optimize the fattening of livestock, whilst minimizing the accumulation of excess internal carcass fat [1, 23, 29, 35, 36] and improving production efficiency. The methods include the manipulation of age, sex and production systems, and the use of biotechnologies such as improved animal feeding, ration formulation, steroidal hormone implants and feed additives such as probiotics, prebiotics and repartitioning agents (beta-adrenergic agonists). The latter molecules repartition energy away from fat accretion to protein (muscle) accretion, by stimulating muscle anabolism and decreasing its catabolism [23, 35]. These molecules are gaining popularity because of the effective repartitioning of energy towards lean gain (muscle growth) and considerable reductions in carcass fat content. Moreover, these molecules improve the use of natural resources through large improvements in animal growth and efficiency, which reduces the environmental impact of animal production systems.

2.2 Fatty acid composition of red meats

Neutral lipids also known as triacylglycerol's which are the major lipid class in animal fats, while phospholipids occur mostly in cell membranes and muscle tissue [20]. Triacylglycerol's serve as a concentrated energy source for the body [36]. Phospholipids contain much higher levels of PUFA and lower SFA content for the subsequent function as a constituent of cellular membranes [20, 31]. Oleic acid (C18:1c-9) is formed from stearic acid (C18:0) by stearoyl CoA desaturase and is a major component of neutral lipids [31]. In ruminant animals, lipids accumulate mainly as triacylglycerol's in adipocytes (fat cells) which are located in subcutaneous, intermuscular and intramuscular adipose tissue depots, as well as abdominal fat depots such as omental and perirenal adipose tissue [36]. Dietary n-6 and n-3 PUFA are incorporated into adipose and muscle tissues despite rumen biohydrogenation. Ruminant animals preferentially incorporate essential fatty acids into muscle tissues rather than into adipose tissue [31].

In ruminant animals, the diet has a small but significant effect on the fatty acids in triacylglycerol's due to rumen biohydrogenation. By contrast, higher proportions of PUFAs are located in the phospholipids than triacylglycerol's, and they are less affected by diet [1]. The PUFAs in phospholipids consist of essential fatty acids such as C18:2n-6 and C18:3n-3 as well as their long-chain derivatives namely ecosapentaenoic acid (C20:5n-3; EPA) and decosahexaenoic acid (C22:6n-3; DHA) [37]. The fatty acid composition determines the firmness and oiliness of adipose tissue as well as the oxidative stability of muscle tissues, which influences the flavor and color of the meat product [24, 25, 31].

The ratio of PUFA/SFA and the n-6/n-3 ratio influence the nutritional value of lipids in red meats. Fats containing a lower SFA content and higher PUFA and CLA content are regarded as ideal [29, 37]. The PUFA/SFA (also referred to as P:S) as well as the n-6/n-3 fatty acid ratios are primary values used to measure the nutritional quality of foods for human nutrition [29]. The recommended benchmark value for the PUFA:SFA (or P:S) ratio is >0.7, and for the n-6/n-3 ratio < 5 [38].

Lean red meat is an excellent source of long-chain omega-3 PUFAs [39]. The PUFA in beef range from 14 to 24%, and C18:1 trans-fatty acid content about 3% of the total fatty acid profile [27, 40]. Conjugated linoleic acid (CLA) occurs naturally in meat and dairy products and is credited for beneficial effects on type II diabetes, weight loss and muscle accretion. CLA's are omega 6-fatty acids which occur in either the cis or trans configuration, with a total of about 28 isomers. CLA's are conjugated fatty acids which means that the two "conjugated" double bonds are not separated by a methylene group, so they are excluded from the TFA's.

2.3 Overview of trans-fatty acids in food

Trans-fatty acids (TFA's) are the sum of all unsaturated fatty acids that contain one or more isolated, non-conjugated double bond in the trans geometric configuration [2]. According to the European Food Safety Authority [3], TFA's are not vital in the human body because they are neither synthesized nor required by the human body. Health professionals worldwide recommend a reduction in the overall consumption of SFA's, TFA's and cholesterol, while increasing the intake of n-3 polyunsaturated fatty acids (PUFAs) [4, 5]. Unfortunately no "Population Reference Intake (PRI)", "Average Requirement (AR)" or "Adequate Intake (AI) value" have been established for TFA's yet [3].

The adverse health effects associated with the consumption of TFA's to date relate specifically to non-communicable diseases such as cardiovascular diseases (CHD and CVD) [3, 6, 7]. It was estimated that CHD is the number one cause of death globally and the consumption of TFA that exceeds five grams per portion is associated with increased risk of CHD [8, 41]. A positive association was confirmed between industrial TFA intake and increased CHD risk [42]. Industrial TFA are produced during partial hydrogenation of vegetable fats and fish oils and are common to products such as margarines, spreads and shortenings, baked goods and deep-fried foods such as those in fast food outlets [11, 43].

There is relatively little information about TFA's in livestock as well as the specific factors that influence TFA concentrations in red meat. The purpose of this section is to present new data about the cis/trans-fatty acid composition of typical beef and lamb meat.

2.4 Effects of production systems on fat and fatty acid composition of red meats

Cattle are frequently fattened on either rangelands or concentrate diets. Different combinations of pasture-based systems together with intensive systems are used to raise livestock and the degree to which these are used rely on resources available as well as climatological conditions [44]. Pasture feeding has gained popularity because of the favorable effects on the fatty acid profile of meat, although the effects are small and at the expense of increased production efficiency [1].

The essential fatty acid alpha-linolenic acid (C18:3n-3) is a prominent fatty acid found in grasses, and therefore a consistent concentration occurs in muscle tissue, despite rumen biohydrogenation [31, 45]. Concentrate fed cattle accumulate slightly less C18:3n-3, which may decrease the n-6/n-3 ratio marginally [18]. A lower n-6/n-3 ratio is obtained when grass and pasture feeding is employed [31]. Grass-fed animals produce meat products that exhibit better oxidative stability due to higher concentrations of natural antioxidants such as vitamin E, which stabilizes the PUFAs [34, 46]. Linoleic acid (C18:2n-6) is a major fatty acid in concentrate feeds (grains and oilseeds). It is degraded into MUFAs and SFAs in the rumen, and incorporated into adipose and muscle tissue in relatively high concentrations [31].

The main benefit of concentrate feeding of livestock is the higher growth rates and feed efficiency achieved, which yield carcasses of high value at a younger age than those from extensive (pasture-based) production systems [44]. This depends typically on better use of feed resources due to meticulous feed formulation and mixing. Concentrate fed cattle generally have better carcass characteristics i.e. a heavier carcass weight, better dressing percentage and conformation scores, and carcass grading or classification compared to steers finished on pasture [45, 47, 48]. Pasture finished cattle yield darker meat with fat which has a more yellow color [45]. The differences in fatty acid content of concentrate fed cattle compared to those raised on pastures are small due to rumen biohydrogenation.

Similar results were reported for lamb and mutton on pasture versus concentrate feeding in France [49, 50] and South Africa [1, 24, 25, 51]. Concentrate fed lambs produced heavier carcasses with better muscular conformation scores and they were fatter than those from grass-fed lambs. The meat from concentrate fed lambs are superior in juiciness as well as in tenderness relative to those from grazing natural pastures. The subcutaneous fat from pasture-fed lambs is yellower and harder, while the meat is darker compared to grain-fed lamb [1, 51]. Meat from grass-fed lambs contain marginally lower triacylglycerol's and a higher phospholipids content [50]. The triglyceride fraction contain marginally higher proportions of C18:0, C18:3n-3, CLA and lower proportions of C16:0, MUFAs, C18:2n-6 and other n-6 PUFAs. The phospholipids contain lower MUFAs, C18:2n-6 and other n-6 PUFAs and higher levels of C18:3n-3. Backgrounding feeding before the onset of concentrate feeding is a relatively new practice. Research indicate that backgrounding strategies have beneficial residual effects on lipid profiles and CLA content of meat from Angus heifers [52].

2.5 Effects of exogenous growth-regulating molecules on fat and fatty acid content of red meats

Red meat producers have been using growth-promoting agents for over 50 years to improve muscle leanness, increase average daily weight gain, stimulate feed intake, and enhance the feed efficiency of animals [35]. Hormonal growth implants and feed additives such as anabolic steroids and beta-adrenergic agonists are used in the beef industry (excluding the European Union) to obtain improved growth rates and feed efficiency during the fattening phase, and subsequently superior carcass composition and quality [44]. The use of these exogenous growth modifiers has improved the effectiveness of red meat production by producing meat that complies more closely with consumer demands in terms of optimizing carcass fat accumulation and thus producing a leaner product [44]. This has resulted in significant improvements in the efficiency of animal production with major benefits to beef producers, retailers and consumers in terms of the relative price competitiveness of beef relative to other dietary protein sources [35, 44].

Beta-adrenergic agonists are compounds similar to naturally occurring endogenous catecholamines (norepinephrine and epinephrine) which are used as feed additives to improve feed efficiency at the end of the fattening phase by placing focus on more protein synthesis rather than fat accretion [23, 35]. These synthetic products such as ractopamine hydrochloric acid and zilpaterol hydrochloric acid provide comparable production benefits as steroid implants but differ in the application as well as mode of action [36]. The results of such treatments during the final 20–25 to 30–42 days of the fattening period are leaner carcasses with improved conformation [23, 35, 53]. The use of beta-adrenergic agonists (L-644,969) improved the proportion and distribution of lean meat [54]. Zilpaterol hydrochloride was shown to decrease carcass fat content and had beneficial effects on fatty acid composition of beef [23, 51]. Higher proportions of oleic acid (C18:1) and lower proportions of C14:0 and C16:0 were deposited in tissues of steers supplemented with zilpaterol hydrochloride, and hence decreasing the saturated fatty acid content of meat [23, 51].

3. Analysis of cis/trans-fatty acid composition of red meats

3.1 Cis/trans fatty acid content of beef and lamb

The purpose of this research was to determine the fat and fatty acid content of beef and mutton, with specific reference to the cis/trans-fatty acid composition. The cis/trans-fatty acid content of South African beef and lamb from range and conventional production systems were analyzed. Representative samples of beef sirloin steaks (n = 60) and lamb loin chops (n = 80) (the lumbar part of the longissimus dorsi muscle) were collected from carcasses from several randomly selected abattoirs in the Gauteng region in South Africa for proximate and fatty acid analyses. The samples were reasonably representative of beef and lamb available from meat retailers in South Africa, with the beef originating from feedlot systems and the lamb from extensively reared lamb production systems.

3.2 Materials and methods

Proximate analysis of beef sirloins and lamb loin chops, as well as medium- and long-chain fatty acid analyses were conducted as previously explained [23, 25]. The fatty acid methyl esters from beef and lamb samples were analyses using gas chromatography. All visible fat was removed from meat samples using a scalpel, followed by mincing and blending of each sample separately into a homogenous mixture using a BUCHI Mixer B-400.

The intramuscular fat content within the muscle tissue was analyzed to determine its nutritional contribution to the human diet. The equipment used to dissect the meat samples (i.e. glass cutting board, scalpel and mixer beaker and blades) were thoroughly cleaned with warm soapy water. All blended meat samples were stored overnight in air-tight containers in the fridge for subsequent proximate analyses and lipid extraction [55]. The remainder of the samples were freeze-dried and then milled for approximately 10 seconds into a fine powder form using a Hamilton Beach Commercial blender. The milled samples were placed into large plastic containers with screw caps and labelled.

Determination of fat content by ether extract was performed using the Soxtec method with the Tecator Soxtec System 1040 Extraction unit, at a temperature setting of 80°C [55]. Ether extraction was done in duplicate by using sub-samples of the freeze-dried, minced meat samples (ca. 2 g). Fatty acid analyses were performed by extracting the lipids from the freeze-dried meat samples using the chloroform: methanol (2:1, v/v) method [56], with modifications as described previously [24]. Saponification and methylation of the fatty acids were done using 14% BF3/CH3OH, followed by analysis with a Varian 3300 FID gas chromatograph fitted with a 100-meter WCOT fused silica capillary column (CP-Sil 88, 100 m x 0.25 mm DF 0.2 μm).

Fatty acids were identified based on the retention times of known fatty acid methyl ester standards obtained from Nu Chek Prep.Inc., Elysim, Minesota (USA). Helium was used as the carrier gas at a flow rate of 50 ml/min, and the gas chromatograph program used was the same as that previously described [24]. Fatty acids were expressed both in terms of the proportion of total medium- and long-chain fatty acids (w/w, %) and in gravimetric concentrations (mg/g of tissue sample) [24, 25].

The raw data was recorded in an excel spreadsheet and all statistical procedures were carried out with the IBM SPSS Statistics Windows software package, version 26 (SPSS Inc., Chicago, IL, USA). Quantitative analysis of the beef and lamb samples were conducted and the means and standard deviations of each lipid fraction, as well as molar proportions (w/w%) and gravimetric content (mg/g meat) of identified fatty acids were determined.

3.3 Composition of cis/trans-fatty acids in beef and lamb

The results of cis/trans-fatty acid analysis of beef and lamb in South Africa are presented in **Tables 2–5** respectively. The results presented in **Table 2** indicate that the fatty acid composition of conventional beef contains about 70% less TFA's (e.g. 1.4 g/100 g beef; **Table 2**) than the recommended threshold of 5 g per 100 g portion, which is the threshold for increased risk of CHD. It should be emphasized that these TFA values were obtained in beef from cattle in conventional production systems in South Africa, i.e. weaner production on extensive grazing, followed by concentrate feeding for about 120 days to the desired carcass composition to meet market requirements, namely carcasses containing about 15–18% carcass fat and 5–7% intramuscular fat (measured in the longissimus dorsi (loin) muscle).

Lipid fraction	Content (%)
IMF	6.5%
SFA	49.61%
MUFA	46.96%
PUFA	3.41%
TFA	4.19%
TFA/100 g meat	1,4 g

MUFA, monounsaturated fatty acids (C18:1 t-11 + C18:1c-9); PUFA, polyunsaturated fatty acids (sum of C18:2 isomers + sum of C18:3 isomers); SFA, saturated fatty acids (C14:0 + C16:0 + C18:0); TFA, trans-fatty acids.

Table 2.
The lipid composition of beef sirloin (Longissimus dorsi samples) from conventional production systems in South Africa (n = 60).

Fatty acid	Molar % (w/w%; n = 60)	Gravimetric content (mg /g meat; n = 60)
C14:0	3.2 ± 0.569	6.36 ± 2.69
C16:0	25.97 ± 3.205	51.67 ± 21.900
C18:0	20.44 ± 5.589	40.01 ± 17.285
C18:1(n-9)t	4.82 ± 1.930	9.32 ± 4.8
C18:1(*n-9*)c	40.71 ± 8.793	83.06 ± 44.641
C18:2 (*n-6*)	3.02 ± 0.955	5.98 ± 3.057
C18:2 (t-9,t-12)	0.14 ± 0.134	0.29 ± 0.206
C18:2 (c-9,t-12)	0.12 ± 0.097	0.27 ± 0.270
C18:2 (t-9,c-12)	0.01 ± 0.029	0.03 ± 0.109
C18:2 (c-9,c-12)	3.75 ± 0.923	7.35 ± 2.724
C18:3 (*n-3*)	0.39 ± 0.183	0.86 ± 0.594
C18:3 (t-9,t-12,t-15)	0.09 ± 0.098	0.20 ± 0.159
C18:3 (c-9,t-12,c-15)	0.07 ± 0.059	0.17 ± 0.179
C18:3 (t-9,c-12,c-15)	0.12 ± 0.094	0.27 ± 0.241
C18:3 (c-9,c-12,c-15)	0.09 ± 0.064	0.22 ± 0.153
CLA (*n-6*)	1.43 ± 0.234	0.90 ± 0.738

CLA, conjugated linoleic acid (C18:2c-9,t-11); MUFA, monounsaturated fatty acids (C18:1 t-11 + C18:1c-9); PUFA, polyunsaturated fatty acids (sum of C18:2 isomers + sum of C18:3 isomers); SFA, saturated fatty acids (C14:0 + C16:0 + C18:0); TFA, trans-fatty acids.

Table 3.
The molar (w/w%) and gravimetric (mg/g) content of cis-and trans-fatty acids in beef sirloin (Longissimus dorsi samples) from conventional production systems in South Africa (w/w%) (n = 60).

Lipid fraction	Content (%)
Intramuscular fat%	13.95 ± 4.18%
SFA	52.13 ± 1.22%
MUFA	38.98 ± 2.65%
PUFA	3.24 ± 0.857%
CLA	0.70 ± 0.148%
TFAs	4.65 ± 1.284%
TFA's/100 g lamb	0.22 g

CLA, conjugated linoleic acid; MUFA, monounsaturated fatty acids (C18:1 t-11 + C18:1c-9); PUFA, polyunsaturated fatty acids (sum of C18:2 isomers + sum of C18:3 isomers); SFA, saturated fatty acids (C14:0 + C16:0 + C18:0); Trans FAs, trans-fatty acids.

Table 4.
The lipid composition of lamb loin chops (Longissimus dorsi samples) from conventional production systems in South Africa (n = 80).

The regression of TFA accumulation with increasing carcass fat content in beef carcasses is presented in **Figure 1** (r2 = 0.64, $P < 0.05$). This illustrates that TFA concentrations accumulate slowly with increasing carcass fat content, but reach a turning point at about ca. 25% carcass fat content. Although the proportions of fatty acids generally remain relatively unchanged in ruminant fats due to rumen biohydrogenation, it seems that the concentration of TFA's reach a turning point

Fatty acid composition	Molar % (w/w%)	Gravimitric content (mg/g meat)
C12:0	0.146 ± 0.057	0.081 ± 0.040
C14:0	3.368 ± 0.673	1.742 ± 0.710
C14:1	0.084 ± 0.024	0.050 ± 0.024
C16:0	24.23 ± 1.788	16.612 ± 4.562
C16:1	1.023 ± 0.192	0.908 ± 0.279
C17:0	1.774 ± 0.525	0.545 ± 0.152
C18:0	22.30 ± 3.321	10.740 ± 2.373
C18:1 (n-11 t)	3.910 ± 1.256	1.735 ± 0.584
C18:1 (n-9c)	33.27 ± 2.809	24.348 ± 5.668
C18:l(n-11c)	0.631 ± 0.114	0.608 ± 0.124
C18:2 (n-6 t)	0.046 ± 0.014	0.019 ± 0.007
C18:2(n-6c)	2.029 ± 0.772	1.918 ± 0.385
C18:2 (n-10 t, n-12c)	0.592 ± 0.133	0.314 ± 0.147
C18:2 (n-10c,n-12c)	0.006 ± 0.001	0.003 ± 0.001
C18:2 (n-9c,n-11 t)	0.013 ± 0.008	0.008 ± 0.006
C18:3(n-6)	0.026 ± 0.005	0.022 ± 0.004
C18:3(n-3)	0.331 ± 0.069	0.176 ± 0.056
C20:0	0.131 ± 0.026	0.052 ± 0.011
C20:1(n-9)	0.060 ± 0.016	0.034 ± 0.006
C20:2	0.029 ± 0.007	0.030 ± 0.006
C22:0	0.011 ± 0.004	0.018 ± 0.004
C20:4(n-6)	0.081 ± 0.020	0.520 ± 0.147

CLA, conjugated linoleic acid (C18:2c-9,t-11); MUFA, monounsaturated fatty acids (C18:1 t-11 + C18:1c-9); PUFA, polyunsaturated fatty acids (sum of C18:2 isomers + sum of C18:3 isomers); SFA, saturated fatty acids (C14:0 + C16:0 + C18:0); Trans FAs, trans-fatty acids.

Table 5.
The molar (w/w%) and gravimetric (mg/g) content of cis-and trans-fatty acids in lamb shops (Longissimus dorsi samples) from conventional production systems in South Africa (w/w%) (n = 80).

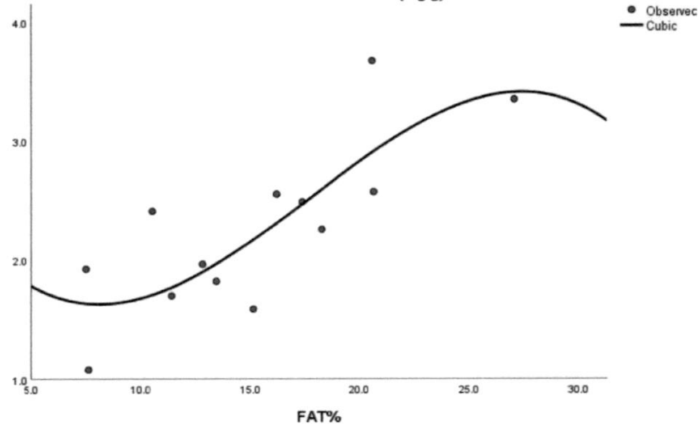

Figure 1.
Regression of trans-fatty acid content of beef loin samples (mg/g) with increasing carcass fat content in conventionally fed cattle.

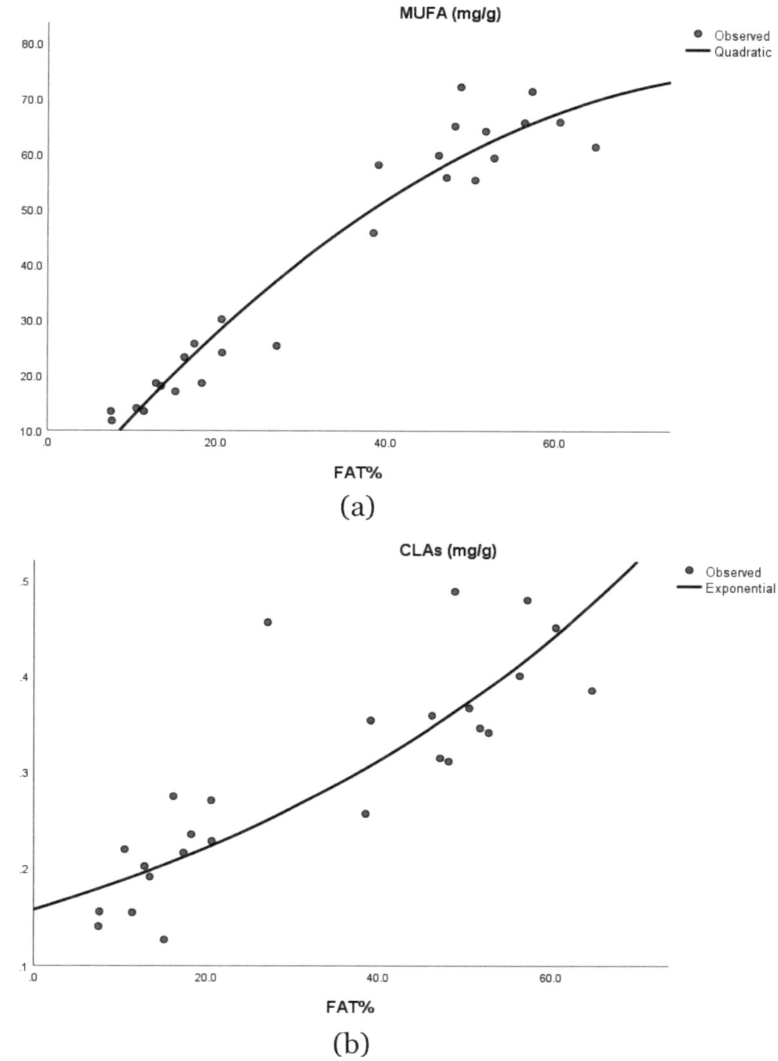

Figure 2.
Regression equations of (a) mono-unsaturated fatty acid content (MUFA, w/w%) of beef loin muscle with increasing carcass fat content, and (b) conjugated linoleic acid content (CLA, w/w%) of beef loin muscle with increasing carcass fat content.

perhaps due to the preferential incorporation other dominant fatty acids such as MUFA and CLA metabolism and absorption in ruminants (see **Figure 2a** and **b**).

The molar composition of fatty acids of beef produced in conventional production systems contain less than 50% saturated fatty acids, while the balance of the fatty acids consist of heart-healthy MUFAs and PUFAs. The trans-fatty acid component of beef is ca. 4.19%, of which a large portion comprise the essential conjugated linoleic acid isomers.

Assessment of the specific medium and long-chain fatty acids in beef (**Table 3**) indicate that the SFAs consist of a small proportion of myristic acid (<3.5% C14:0), about 25% of palmitic acid (C16:0) and 20% of stearic acid (C18:0). It follows that the medium-chain fatty acids (MCFA; C14:0 and C16:0) comprise less than a third of the fatty acid content of beef, which is beneficial from a heart health perspective, as MCFA's have been associated with a slight elevation in low-density lipoprotein

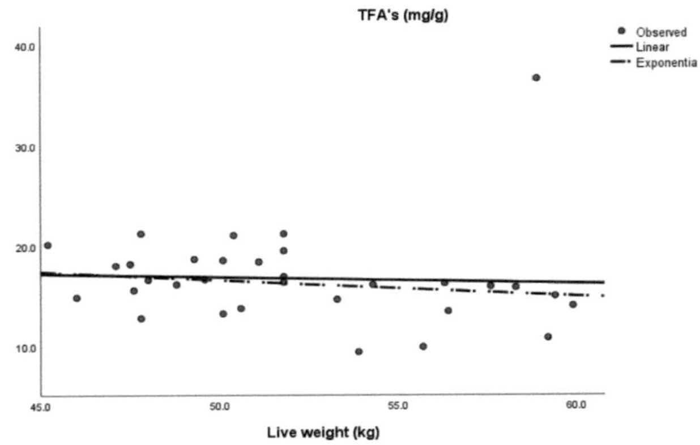

Figure 3.
Trans-fatty acid content (TFA, mg/g) of lamb loin muscle with increasing live weight.

(LDL) cholesterol values [57]. Stearic acid is known as a cholesterol neutral fatty acid and therefore beneficial in terms of essential fat ingestion. Recent research indicate that stearic acid decreases both cardiac and cancer risk in humans [58]. This beneficial fatty acid is also credited with signaling the mitochondria in cells and stimulate fatty acid beta-oxidation. The MUFAs in beef consist predominantly (40%) of the LDL cholesterol lowering oleic acid (C18:1) which therefore has no adverse effects on CHD or CVD. A large portion of the PUFAs also consist of CLA isomers, which are known to occur predominantly in milk and meat from ruminant livestock such as cattle and sheep.

Oleic acid (C18:1) has a well-known reputation of lowering the risk of cardiovascular disease [59]. In this review, it is emphasized that the replacement of 5% of SFA's with oleic acid, may reduce the risk of coronary heart disease by 20–40%. The C18:1-component of beef is almost 45%, and it is important to know that concentrate feeding and dietary supplementation of cattle diets with zilpaterol hydrochloride increases the oleic acid content of beef by ca. 2% [23].

Figure 2(a) indicates the highly significant correlation between the accumulation of MUFA's in beef loin muscle with increasing carcass fat percentage ($r^2 = 0.94$, $P < 0.001$). Similarly, the CLA content of beef loin muscle (mg/g of loin muscle) correlates significantly ($r^2 = 0.709$, $P < 0.001$) with increasing carcass fat content in beef carcasses, as illustrated in **Figure 2(b)**.

This implies that the accumulation of both MUFA's and CLA's in beef muscle is significantly influenced by the increasing maturity and fat accumulation in cattle, in addition to the known dietary effects on beef CLA's. Beef carcasses are generally fed to about 18–25% carcass fat in European countries, Australia and South Africa and to about 25 to 35% carcass fat in the USA. Backgrounding on pastures combined with intensive concentrate feeding have beneficial effects on the CLA content of beef.

Lamb meat contains more intramuscular fat (ca. 14%; **Table 4**) compared to beef. However, the fat is not all saturated as commonly believed, with SFA's seldom exceeding 53%, and about 40% MUFA's, 3,2% PUFA's and low trans-fatty content (**Tables 4** and **5**). The data from typical mutton sheep breeds raised mainly on extensive pasture systems in South Africa indicate a low TFA content of 0.22 g/100 g meat, which is much lower than the threshold value of 5 g / 100 g portion.

It has been postulated that ruminant derived trans vaccenic acid may not be associated with an increased risk of CHD or CVD, because it is converted to CLA isomers, which hold several health benefits [14, 15]. Vaccenic acid in adipose tissues is converted to CLA by the stearoyl-CoA desaturase enzyme [31]. This enzyme is also responsible for converting oleic acid (C18:1c-9) to stearic acid (C18:0).

Figure 3 confirms that the TFA content of lamb loin muscle remains consistently low during the normal popular target weights at slaughter namely between 40 to 55 kg live weight.

4. Discussion

It is accepted that the consumption of excess red meat and alcohol, may adversely affect the life expectancy of humans [60]. Nonetheless, almost all lipid-containing foods in Western diets contribute to n-6 PUFA and alpha-linolenic acid intake, while only meat, eggs and seafood contribute to beneficial n-3 PUFA intake in the human diet [61]. Results of published epidemiological studies are conflicting concerning the effects of TFA's on blood cholesterol homeostasis [53, 62–66]. Unfortunately, most dietary trials on TFA's are based on industrial TFA's with the assumption that those trans isomers have the same metabolic effects as TFA's from natural sources such as those from red meat. The latter assumption is incorrect because trans vaccenic acid (C18:1 t-11) is unique of TFA's in ruminant adipose tissue, while elaidic acid (C18:1 t-9) occurs specifically in industrial TFA's [68].

The most important trans isomers in human biology are the mono-unsaturated and di-unsaturated fatty acids which contain sixteen and eighteen carbon atoms, namely C16:1 trans, C18:1 trans and C18:2 trans-fatty acids [67]. C18:1 trans isomers encompass 80–90% of total TFA content in foods. The importance of different tertiary structures relay to the different crystalline packing which ultimately result in differing melting points of these fatty acids [67]. The melting points of the trans isomeric fatty acids are generally much higher compared to their corresponding cis-isomers, i.e. oleic acid (C18:1 cis) has a melting point of 10–11°C, while its trans-isomer, elaidic acid (C18:1 trans) has a melting point of 44.5–45.5°C [67].

Based on the data from the present study, the TFA content of beef and lamb is low, and tend to peak early in the growth curve, after which the relative concentration starts to decrease slowly. This is associated with a slow but consistent increase in CLA's in ruminant muscle tissues. The current data also indicate that the growth curve and carcass fat content affect the accumulation of TFA's and CLA's in addition to the known effects of nutrition on their accumulation. Conjugated linoleic acid (CLA) is not included in the definition of TFA as it has conjugated double bonds [3], which means that the two double bonds are "conjugated" or continuous or not separated by a methylene group. CLA is a collective term used for all conjugated geometric and positional isomers of C18:2 (linoleic acid) [68, 69]. It is less well known that meat and milk from ruminant livestock are excellent sources of CLA's, which are synthesized in the rumen by the microbial isomerisation and biohydrogenation of mostly linoleic and alpha-linolenic acids (e.g. 18 carbon PUFAs), and the desaturation of trans-fatty acids in their adipose tissue and mammary glands. Rumenic acid (C18:2 c9,t11) comprises about 90% of the total CLA isomer range [70]. More recently the CLA-isomer (C18:2 t7,c9) has been detected in milk, cheese, beef, human milk as well as human adipose tissue.

CLA's are nutritionally beneficial due to their antioxidant and anti-carcinogenic properties [70, 71]. CLA's also have beneficial effects on non-communicable diseases such as cancer, cardiovascular diseases, diabetes, and positive effects on the immune system and bone health [70].

5. Conclusions

Red meats constitute an important source of high-quality protein and fatty acids in the human diet. Accordingly, red meat forms a critical part of a balanced diet, along with moderate food intake and healthy habits such as physical activity and not smoking. The intramuscular fat content of red meats range between 3 and 14%, which is low compared to most fat-containing processed foods (25%). More importantly, the fats in red meats are not all saturated, with a SFA content of 45 to 53%, consisting predominantly of stearic acid which has proven health benefits. Animal fats contain several other beneficial fatty acids (including MUFA's, PUFA's, and CLA's) that have either no, or an LDL-cholesterol lowering effect, which may reduce the risk of coronary heart diseases.

TFA's are neither synthesized nor required by the human body. Although health professionals recommend a reduction in the consumption of SFA and TFA, animal fats contribute a very small portion of the overall TFA intake, most of which originate from other sources such as the partial hydrogenation of vegetable and fish oils. The lean component of beef and lamb contains a trivial proportion of TFA's, consisting of vaccenic acid, rumenic acid and CLA-isomers. CLA's in red meat are beneficial due to their antioxidant and anti-carcinogenic properties, so they should not strictly be considered in the TFA definition. This means that the TFA's in red meats are negligible and pose no harm to human health. Better labelling of red meats is recommended, to ensure that consumers understand the nutritional value of red meats, as part of a balanced diet.

Acknowledgements

The research support of postgraduate students Belinda de Jong, Andrea Hasewinkel and Erna Mostert are acknowledged. The advice and recommendations of Dr. Elizabeth M. Webb from the School of Health Systems and Public Health, Faculty of Health Sciences at the University of Pretoria are acknowledged.

Conflict of interest

There is no conflict of interest.

Author details

Edward C. Webb
Department of Animal Science, Production Animal Physiology and Meat Science Research Group, University of Pretoria, Hatfield, South Africa

*Address all correspondence to: edward.webb@up.ac.za

References

[1] Webb , E.C. & O'Neill, H.A. The animal fat paradox and meat quality, Meat Science, 2008; 80, 28-36.

[2] Albuquerquea, T.G., Costa, H.S., Castilho, M.C. & Sanches-Silva, A., Trends in the analytical methods for the determination of trans fatty acids content in foods. Trends in Food Science & Technology. 2011; 22, 543-560.

[3] EFSA Panel on Dietetic Products, Nutrition, and Allergies (NDA)., 2010. Scientific Opinion on Dietary Reference Values for fats, including saturated fatty acids, polyunsaturated fatty acids, monounsaturated fatty acids, trans fatty acids, and cholesterol. EFSA Journal. 8, 1-107.

[4] Griel, A.E. & Kris-Etherton, P.M. Beyond saturated fat: The importance of the dietary fatty acid profile on cardiovascular disease. Nutrition Reviews. 2006; 64, 257-262.

[5] Kris-Etherton, P.M. & Innis, S. Dietary Fatty Acids – Position of the American Dietetic Association and Dietitians of Canada. Journal of the American Dietetic Association. 2007; 107, 1599-1611.

[6] Judd, J.T., Clevidence, B.A., Muesing, R.A., Wittes, J., Sunkin, M.E. & Podczasy, J.J. Dietary trans fatty acids: Effects on plasma lipids and lipoproteins of healthy men and women. American Journal of Clinical Nutrition. 1994; 59, 861-868.

[7] Mensink, R.P. & Katan, M.B. Effect of dietary trans fatty acids on high-density and low-density lipoprotein cholesterol levels in healthy subjects. The New England Journal of Medicine. 1990; 23, 439-445.

[8] Wagner, K., Plasser, E., Proell, C. & Kanzler, S. Comprehensive studies on the trans fatty acid content of Austrian foods: Convenience products, fast food and fats. Food Chemistry. 2008; 108, 1054-1060.

[9] Ratnayake, W.M. & Zehaluk, C. Trans fatty acids in foods and their labeling regulations. In: Healthful Lipids. Eds Akoh, C.C. & Lai, O.M., AOCS Press, 2005; pp 1-32.

[10] Richter, E.K., Shawish, K.A., Scheeder, M.R. & Colombani, P.C. Trans fatty acid content of selected Swiss foods: the TransSwissPilot study. Journal of Food Composition and Analysis. 2009; 22, 479-484.

[11] Stender, S., Astrup, A. & Dyerberg, J. Ruminant and industrially produced trans fatty acids: health aspects. 2008; Food & Nutrition Research.

[12] Fritsche, J. & Steinhart, H. Contents of trans fatty acids (TFA) in German foods and estimation of daily intake. Lipid. 1997; 99, 214-217.

[13] Weggemans, R.M., Rudrum, M. & Trautwein, E.A. Intake of ruminant versus industrial trans fatty acids and risk of coronary heart disease – what is the evidence? European Journal of Lipid Science and Technology. 2004; 106, 390-397.

[14] Huth, P.J. Do ruminant trans fatty acids impact coronary heart disease risk? Lipid Technology. 2007; 19, 59-62.

[15] Turpeinen, A.M., Mutanen, M., Aro, A., Salminen, I., Basu, S., Palmquist, D.L. & Griinari, J.M. Bioconversion of vaccenic acid to conjugated linoleic acid in humans. American Journal of Clinical Nutrition. 2002; 76, 504-510.

[16] Zock, P.L. & Mensink, R.P. Dietary trans fatty acids and serum lipoproteins in humans. Current Opinions of Lipidology.1996; 7, 34-37.

[17] Katan, M.B. Trans fatty acids and plasma lipoproteins. Nutrition Reviews. 2000; 58, 188-191.

[18] Nuernberg, K., Dannenberger, D., Nuernberg, G., Ender, K., Voigt, J., Scollan, N.D., Wood, J.D., Nute, G.R. & Richardson, R.I. Effect of grass-based and a concentrate feeding system on meat quality characteristics and fatty acid composition of longissimus muscle in different cattle breeds. Livest. Prod. Sci. 2005; 94, 137-147.

[19] De Smet, S., Webb, E. C., Claeys, E., Uytterhaegen, L., & Demeyer, D. I. Effect of dietary energy and protein levels on fatty acid composition of intramuscular fat in double-muscled Belgian Blue bulls. Meat Science, 2000; 56, 73-79.

[20] Webb, E. C., De Smet, S., Van Nevel, C., Martens, B., & Demeyer, D. I. Effects of anatomical location on the composition of fatty acids in double-muscled Belgian Blue cows. Meat Science, 1998; 50, 45-53.

[21] Enser, M., Hallett, K. G., Hewett, B., Fursey, G. A. J., Wood, J. D., & Harrington, G. Fatty acid content and composition of UK beef and lamb muscle in relation to production system and implications for human nutrition. Meat Science, 1998; 49, 329-341.

[22] Enser, M., Hallett, K., Hewitt, B., Fursey, G. A. J., & Wood, J. D. Fatty acid content and composition of English beef, lamb and pork at retail. Meat Science, 1996; 42(4), 443-456.

[23] Webb, E. C., & Casey, N. H. Fatty acids in carcass fat of steers treated with ß-adrenergic agonist individually or in combination with trenbolone acetate + oestradiol-17b. Meat Science, 1995; 41(1), 69-76.

[24] Webb, E. C., & Casey, N. H. Genetic differences in fatty acid composition of subcutaneous adipose tissue in Dorper and SA Mutton Merino wethers at different live weights. Small Ruminant Research, 1995; 18, 81-88.

[25] Webb, E. C., Casey, N. H., & Van Niekerk, W. A. Fatty acids in the subcutaneous adipose tissue of intensively fed SA Mutton Merino and Dorper wethers. Meat Science, 1994; 38, 123-131.

[26] Weaver, C.M., Dwyer, J.M., Fulgoni, V.L., King, J.C., Leveille, G.A., MacDonald, R.S., Ordovas, J & Schnakenberg, D., Processed foods: contributions to nutrition, Am J Clin Nutr 2014;99:1525-42.

[27] Williams, P. Key nutrients delivered by red meat in the diet. Nutrition & Dietetics. 2007; 64,113-119.

[28] Wu, G. Important roles of dietary taurine, creatinine, anserine and 4-hydroxyproline in human nutrition and health, Amino Acids, 2020; 52, 329-360.

[29] Warren, H.E., Scollan, N.D., Enser, M., Hughes, S.I., Richardson, R.I. & Wood, J.D. Effects of breed and a concentrate or grass silage diet on beef quality in cattle of 3 ages. I: Animal performance, carcass quality and muscle fatty acid composition. Meat Science. 2008; 78, 256-269.

[30] Droulez, V., William, P.G., Levy, G., Stobaus, T. & Sinclair, A. Composition of Australian red meat, Fatty acid profile. Food Australia. 2006; 58, 335-341.

[31] Wood, J. D., Richardson, R. I., Nute, G. R., Fisher, A. V., Campo, M. M., Kasapidou, E., et al. Effects of fatty acids on meat quality: A review. Meat Science, 2008; 78,343-358.

[32] Tshabalala, P. A., Strydom, P. E., Webb, E. C., & De Kock, H. L. Meat quality of designated South African indigenous goat and sheep breeds. Meat Science, 2003; 65, 563-570.

[33] Mostert, E. The effect of Karoo browse and veldt feeding on long-chain and volatile fatty acid components in lamb, [thesis]. University of Pretoria, 2020.

[34] Laaksonen, D. E., Nyyssonen, K., Niskanen, L., Rissanen & Salonen, J. T. Prediction of cardiovascular mortality in middle-aged men by dietary and serum linoleic and polyunsaturated fatty acids. Archives of Internal Medicine, 2005; 165, 193-199.

[35] Johnson, B.J., Ribeiro, F.J.B. & Beckett, J.L. Application of growth technologies in enhancing food security and sustainability. Anim. Frontiers 2013; 3, 8-13.

[36] Scollan, N. D., Choi1, N.-G., Kurt, E., Fisher, A. V., Enser, M., & Wood, J. D. Manipulating the fatty acid composition of muscle and adipose tissue in beef cattle. Meat Science, 2001; 85, 115-124.

[37] Zervas, G. & Tsiplakou, E. The effect of feeding systems on the characteristics of products from small ruminants. Small Ruminant Research 2011; 101, 140– 149.

[38] Raes, K., De Smet, S., & Demeyer, D. Effect of dietary fatty acids on incorporation of long chain polyunsaturated fatty acids and conjugated linoleic acid in lamb, beef and pork meat: A review. Animal Feed Science and Technology, 2004; 113, 199-221.

[39] National Health and Medical Research Council., 2006. Nutrient Reference Values for Australia and New Zealand Including Recommended Dietary Intakes. Canberra: Commonwealth Department of Health and Ageing.

[40] Motard-Bélanger, A., Charest, A., Grenier, G., Paquin, P., Chouinard, Y., Lemieux, S., Couture, P. & Lamarche, B. Study of the effect of trans fatty acids from ruminants on blood lipids and other risk factors for cardiovascular disease. American Journal of Clinical Nutrition. 2008; 87,593-9.

[41] World Health Organization (WHO). 2008-2013 action plan for the global strategy for the prevention and control of noncommunicable diseases: Prevent and control cardiovascular diseases, cancers, chronic respiratory diseases and diabetes. Geneva: Switzerland. 2008.

[42] Willett, W.C., Stampfer, M.J., Manson, J.E., Colditz, G.A., Speizer, F.E., Rosner, B.A., Sampson, L.A. & Hennekens, C.H. Intake of trans fatty acids and risk of coronary heart disease among women. Lancet. 1993; 341, 581-585.

[43] Ascherio, A., Katan, M.B., Stampfer, M.J. & Willett, W. C. Trans fatty acids and coronary heart disease. The New England Journal of Medicine. 1999; 340, 1994-1998.

[44] Webb, E.C. & Erasmus, L.J., (2013). The effect of production system and management practices on the quality of meat products from ruminant livestock, S. Afr. J. Anim. Sci., 43(3):413-423.

[45] Realini, C.E., Duckett, S.K., Brito, G.W., Dalla Rizza, M. & De Mattos, D. Effect of pasture vs. concentrate feeding with or D. Without antioxidants on carcass characteristics, fatty acid composition, and quality of Uruguayan beef. Meat Science. 2004; 66, 567-577.

[46] Gatellier, P., Mercier, Y. & Renerre, M. Effect of diet finishing mode (pasture or mixed diet) on antioxidant status of Charolais bovine meat. Meat Science. 2004; 67, 385-394.

[47] Keane, M.G. & Allen, P. Effects of production system intensity on performance, carcass composition and

meat quality of beef cattle. Livestock Production Science. 1998; 56, 203-214.

[48] Muir, P.D., Beaker, J.M. & Brown, M.D. Effects of forage- and grain-based feeding systems on beef quality: A review. N. Z. J. Agric. Res. 1998; 41, 623-635.

[49] Priolo, A., Micol, D., Agabriel, J., Prache, S. & Dransfield, E. Effect of grass or concentrate feeding system on lamb carcass and meat quality. Meat Sci. 2002; 62, 179-185.

[50] Aurousseau, B., Bauchart, D., Calichon, E., Micol, D. & Priolo, A. Effect of grass or concentrate feeding systems and rate of growth on triglyceride and phospholipid and fatty acids in the *M. longissimus* thoracis of lambs. Meat Sci. 2004; 66, 531-541.

[51] Webb, E. C., & Casey, N. H. (1997). Influence of dietary presentation on the composition of fatty acids and sensory characteristics of meat from wethers. South African Journal of Food Science and Nutrition, 9, 69-76.

[52] Pordomingo, A.J., Garcia, T.P. & Volpi Lagreca, G. Effect of feeding treatment during the backgrounding phase of beef production from pasture on: II. Longissimus muscle proximate composition, cholesterol and fatty acids. Meat Science. 2012; 90, 947-955.

[53] Strydom, P.E., Hope-Jones, M, Frylinck, L. & Webb, E.C. The effects of a beta-agonist treatment, vitamin D3 supplementation and electrical stimulation on meat quality of feedlot steers, 2011; Meat Sci., 89:462-468.

[54] Moloney, A.P., Allen, P., Ross, D.B., Olson, G, & Convey, E.M. Growth, feed efficiency and carcass composition of finishing Friesian steers fed the beta-adrenergic agonist L-644,969. Journal of Animal Science. 1990; 68, 1269-1277.

[55] AOAC, Official methods of analysis of AOAC International by Patricia A Cunniff, AOAC International. Association of Official Analytical Chemists, Washington, DC, USA, 1995.

[56] Folch, J., Lees, M. & Stanley, H.S. A simple method for the isolation and purification of total lipids from animal tissues. Journal of Biological Chemistry. 1957; 226, 497-509.

[57] Temme, EH, Mensink, RP & Hornstra, G. Effects of medium chain fatty acids (MCFA), myristic acid, and oleic acid on serum lipoproteins in healthy subjects, J Lipid Res., 1997; 38(9):1746-1754

[58] Senyilmaz-Tiebe, D, Pfaff, D.H., Virtue, S, Schwarz, K.V., Fleming, T, Altamura,S, Muckenthaler, M.U., Okun,J.G. Vidal-Puig, A. Nawroth, P. & Teleman, A.E., Dietary stearic acid regulates mitochondria in vivo in humans, Nature Communication, DOI: 10.1038/s41467-018-05614-6

[59] Lopez-Huertas, E. Health effects of oleic acid and long chain omega-3 fatty acids (EPA and DHA) enriched milks. A review of intervention studies, Pharmacological Research 2010; 61, 200-207, DOI: 10.1016/j.phrs.2009.10.007

[60] Chhabi Lal Ranabhat, Myung-Bae Park & Chun-Bae Kim, Influence of alcohol and red meat consumption on life expectancy: Results of 164 Countries from 1992 to 2013, Nutrients, 2020; 12, 459; doi:10.3390/nu12020459.

[61] Meyer, B.J., Mann, N.J., Lewis, J.L., Milligan, G.C., Sinclair, A.J. & Howe, P.R.C. Dietary intakes and food sources of omega-6 and omega-3 polyunsaturated fatty acids. Lipids. 2003; 38, 391-397.

[62] Ascherio, A., Katan, M.B., Stampfer, M.J. & Willett, W. C. Trans fatty acids and coronary heart disease. The New

England Journal of Medicine. 1999; 340, 1994-1998.

[63] Aro, A. Complexity of issue of dietary trans fatty acids. The Lancet. 2001; 357, 732.

[64] Clifton, P.M., Keogh, J.B. & Noakes, M. Trans fatty acids in adipose tissue and the food supply are associated with myocardial infarction. Journal of Nutrition. 2004; 134, 874-879.

[65] Oomen, C.M., Ocke, M.C., Feskens, E.J., Erp-Baart, M.A., Kok, F.J. & Kromhout, D. Association between trans fatty acid intake and 10-year risk of coronary heart disease in the Zutphen Elderly Study: a prospective population-based study. Lancet. 2001; 357, 746 –751.

[66] Sun, Q., Ma, J., Campos, H., Hankinson, S.E., Manson, J.E., Stampfer, M.J., Rexrode, K.M., Willett, W.C. & Hu, F.B. A prospective study of trans fatty acids in erythrocytes and risk of coronary heart disease. Circulation. 2007; 115, 1858-1865.

[67] Aro, A., Kosmeijer-Schuil, T., van de Bovenkamp, P., Hulshof, P., Zock, P. & Katan, M.B., 1998. Analysis of C18:1 cis and trans fatty acid isomers by the combination of gas-liquid chromatography of 4,4-dimethyloxazolinederivatives and methyl esters. Journal of American Oil Chemistry Society. 75, 977-985.

[68] Koletzko, B & Decsi, T. Metabolic aspects of trans fatty acids. Clinical Nutrition. 1997; 16, 229-237.

[69] Schmid, A., Collomb, M., Sieber, R. & Bee, G. Conjugated linoleic acid in meat and meat products: A review. Meat Science. 2006; 73, 29-41.

[70] Yurawecza, M.P., Roacha, J.A.G., Sehata, N., Mossobaa, M.M., Kramerb, J.K.G., Fritschec, J., Steinhart, H. & Kua, Youh. A New Conjugated Linoleic Acid Isomer, 7 trans, 9 cis-Octadecadienoic Acid, in Cow Milk, Cheese, Beef and Human Milk and Adipose Tissue, Lipids. 1998; 33, 803-809.

[71] McDonald, H.B. Conjugated linoleic acid and disease preventions: A review of current knowledge. The American College of Nutrition. 2000; 19, 111-118.

Section 3

Meat Consumption and Handling

Chapter 8

Preservation of Seafoods by Hurdle Technology

Jag Pal, Om Pravesh Kumar Ravi, Sangeeta Kumari and Akhilesh Kumar Singh

Abstract

Fish and seafoods are a highly perishable product due to the biochemical composition and the high microbial load on the skin and gills of fish. The natural microflora that is more adopted to low temperatures results in lower thermal bacterial shock at the storage temperature. The development of new fish processing techniques is required because the demand for fish or seafood with minimum changes in sensory Biochemical and nutritional quality. This has led to the advent of hurdle technology in the field of seafood technology. Hurdle technology is the combined use of several preservation methods to make a product shelf-stable, to improve quality and to provide additional safety. This technology is used in many countries of the world, including India. The factors used for food preservation is called as hurdles. A combination of hurdle such as high temperature, refrigeration, irradiation, drying and smoking etc. are applied to eliminate the growth of microorganism. The application of several hurdle may reduce the rate of fish spoilage caused by spoilage microbe. The objective of this book chapters is focus on the preservative effect of hurdles technology on the quality and shelf-life fish and fishery products with recent, combined updated information.

Keywords: hurdle technology, seafood, quality, spoilage and microorganism

1. Introduction

Since the time immemorial, human society has been dependent on foods for their existence. We derive a variety of foods from plants and animals such as sea-foods, eggs, fishes etc. All these are consumed by man to satisfy their nutritional requirements, proper body growth and development, health and to increase their appetite value. But the food items of animal origin are perishable in nature as a result they get spoiled due to microbial activity. Hence to reduce this toxicity of food 'hurdle technology' was developed several years ago for the production of safe, stable, nutritious, tasty and economical foods. The hurdle technology, also called combined methods or combined processes, is an integrated approach of basic food preservation methods for making the food more safe, stable and nutritious [1]. It can also be defined as a method of achieving control or elimination of pathogens present in the foods for creating safe and shelf - stable food. The concept of hurdle technology is quite old but has been used by many countries across the globe including India for effective preservation of foods. It advocates the wise use of a combination of different preservation factors or techniques termed as hurdles in order to achieve multi-target, mild but reliable preservation effects in foods. Hurdle technology developed with the concept to address the consumer demand to provide more natural and fresh food. According to Leistner [2] hurdle technology is the process of an intelligent combination of hurdles which safeguards the microbial safety and storage stability along with retains the sensory, nutritional quality and economic viability of food materials. It has come in response to several number of developments viz.,

i. To fulfill consumers' demand for improved foods that retain their unique nutritional properties with freshness.

ii. This technology shift food products ready-to-eat and makes it convenience foods with lesser processing requirements by consumers.

iii. Basically, consumers prefer more 'natural food products' which requires less processing effort with use of minimum chemical preservatives.

It offers a framework for merging a number of milder preservation techniques to attain an improved level of product safety with longer stability (dimorianreview.com).

The hurdles technology can be applied in several fish and fishery products such as the salted fish, smoked, marinated products, pickles, canned fish products (high or low temperature), traditional Asian sauces (fermentative microorganisms) and more recently, in vacuum-packed products (redox potential). These preservative factors have been studied for several years ago. Hurdles technology have some prose (by inhibiting microorganisms) along with some cons (on other parameters such as nutritional properties or sensory quality), depending on their intensity. For example, salt content in food must be in such a limit to inhibit pathogens and spoilage microorganisms, but not too high to impair taste and act as pro-oxidants [3].

2. Significance of hurdle technology

1. Improves foodstuffs quality and ensures microbial activity.

2. Food remains safe and stable for a longer duration of time.

3. It is high in sensory and nutritive value due to hurdle effects.

4. It is applicable in both small- and large-scale industries.

5. Does not change the composition or integrity of food items.

6. It fulfills the current demands of the consumers for fresh, natural and minimally processed foodstuffs.

7. It is economically suitable for the nation as it saves money, time, energy and other resources.

3. History of hurdle technology

Food preservation techniques are an inseparable part of production of foodstuffs in order to overcome or counter the pathogen activities. Thus, the food scientist applied combined use of several preservation methods including physical or chemical which is an age-old practice. This is generally used by the food industry to ensure food safety and stability. Seafood in terms of fresh fish which is a highly perishable product due to its chemical composition and high accumulation of microorganisms on its body surface. In 1976, Leistner introduced the term "Hurdle effect". Leistner and co-workers acknowledged that the hurdle concept illustrates the well - known fact that the complex interactions of temperature, water activity, pH, redox potential, etc. are significant for the microbial stability of foods [4].

4. Principle of hurdle technology

Hurdle is defined as a factor, a condition, or a processing step that limits, or prevents the microbial growth and reduce microbial load. There are many preservation methods used for making foods stable and safe e.g., heating, chilling, freezing, freeze drying, drying, curing, salting, sugar addition, acidification, fermentation, smoking and oxygen removal. Currently, more than 50 hurdles are used in food processing industries throughout the world. Some of the principle hurdles used for seafood products are given in **Table 1**. The hurdle effect is the most fundamental importance for the preservation of foods, since the hurdles in a stable product, control the microbial spoilage, food poisoning and desired fermentation processes [5]. If the intensity of a particular hurdle in a food is too small, it should be strengthened, and in case, it is detrimental to the food quality.

4.1 The principle behind this hurdle technology can be summarized in the following ways

1. The preservative factors or hurdles disturbs the homeostatis of microorganisms.

2. Pathogens should not be allowed to cross or jump over all the hurdle effects present in the food items.

3. Preservative factors should not allow the microorganisms to proliferate and remain in an inactive stage or even kill them.

4. The hurdle effect shows that the complexity of interactions of temperature, water, pH, humidity are important factors to microbial stability.

Parameter	Symbol	Application
High temperature	F	Heating
Low temperature	T	Chilling, freezing
Reduced water activity	a_w	Drying, curing, conserving
Increased acidity	pH	Acid addition or formation
Reduced redox potential	Eh	Removal of oxygen or addition of ascorbate
Bio preservatives	—	Competitive flora such as microbial fermentation
Other preservatives	—	Sorbates, sulphites, nitrites

Table 1.
Principle hurdles used for food preservation.

5. Basic aspects of hurdle technology

The hurdle technology affects the physiology and growth of microorganisms in food. There are mainly 4 major mechanisms by which hurdle technology affects the growth of microorganisms in foods, these are *-Homeostasis, Metabolic exhaustion, stress reaction and Multi target preservation* (**Figure 1**).

5.1 Homeostasis

The literal meaning of "homeostasis" is "same state" and it refers to the process of keeping the internal body environment in a steady state, when the external environment is changed. In case of food preservation techniques, if somehow the homeostasis of the pathogens are disturbed, then they will not be able to proliferate themselves. They will remain in the lag phase or even die until their internal body temperature is maintained or recovered [7]. But microorganisms can acquire myriad routes to reach their homeostatic state. Thus, the most effective way to prevent their growth on food items is to go through combined methods of hurdle effect. Disturbing the homeostasis of the microorganisms by various hurdles eventually results in the death of the spoilage causing microbes thereby protecting the food product from microbial spoilage.

5.2 Metabolic exhaustion

Auto-sterilization of food products can be achieved by metabolic exhaustion, which leads to the death of the germinated spores and thus ensuring the success

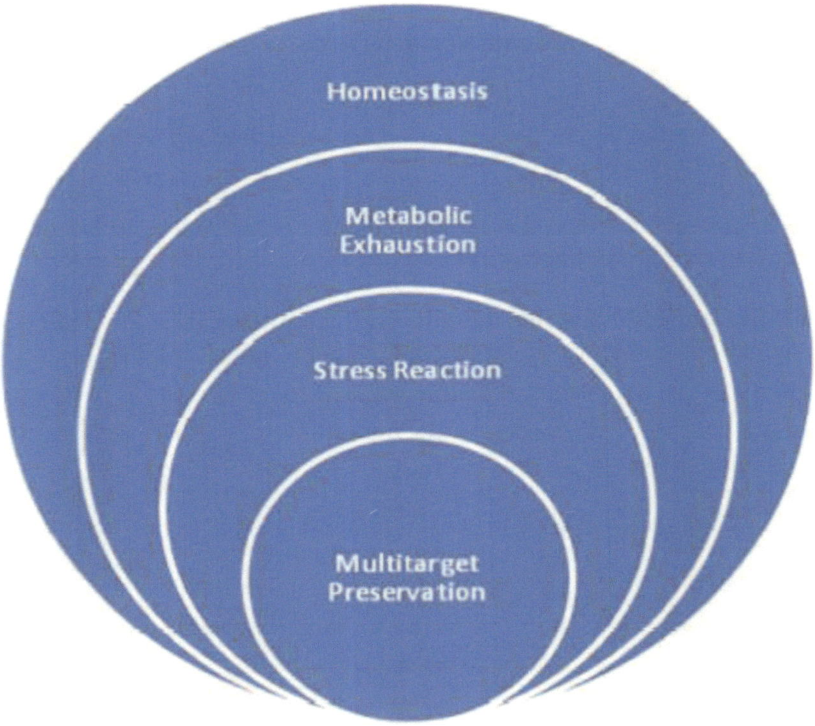

Figure 1.
Basic aspects of hurdle technology [6].

of hurdle technology. There are a number of different types of bacteria, mold and yeast which overcome and sustain the high temperature. Many bacterial spores which survive the thermal treatment are able to germinate in similar food products under unfavorable conditions than those under which vegetative bacteria are able to multiply [8]. Therefore, the microorganisms in the food products try every possible way to repair mechanisms for their homeostasis. By doing this, they completely use up their energy and die. This leads to auto-sterilization of foods. When multiple hurdles are applied in the food items, the rate of metabolic exhaustion is accelerated. In this due course, high energy is required by the microorganisms to maintain their homeostasis which is not achieved by them. Thus, it leads to microbial cell damage and inhibits their further growth [9].

5.3 Stress reactions

Due to generation of shock proteins, some bacteria become more and more virulent under stress conditions. The stress shock proteins are a family of proteins that are produced by the cells in response to exposure to stressful conditions, induced by heat, pH, a_w, ethanol, oxidative compounds, cold, UV light and starvation. Then, the simultaneous exposure to different stresses of bacteria will require high energy demand or at least much more protective stress shock proteins, which ultimately causes the death of the microorganisms [10]. Exposure to the multiple stresses simultaneously induces energy utilizing and synthesis of several stress shock proteins, in turn making the microbes metabolically weak. Hence, multitarget preservation of foods could be an efficient approach for minimizing the production of stress shock proteins and in food preservation for long term [2].

6. Multi target preservation of food

Leistner [11] has been developed the concept of "Multi-target preservation of food" which is a most significant aspect for efficient and effective preservation of targeted food products. Hurdles which are applied in the targeted food products might not just have effects on microbial stability but also it act synergistically [11]. The synergistic effect could be attained in the targeted food products, if the combined effects of different hurdles viz., pH, a_w, Eh, enzyme systems targets simultaneously within the microbial cell and thus disturb the homeostasis of the microbes. This phenomenon interpreting and difficult for the microbes to synthesize a number several stress shock proteins and to maintain their homeostasis [4]. Hence the application of multiple hurdles technique simultaneously would lead to an optimal microbial stability and effective food preservation.

7. Objectives hurdle technology

The main objective of this technology is food preservation, storage of food products and enhancement of their shelf life thereby giving us good quality products. There are several reasons for preserving the foods which are as follow (i) To ensure the safety of food from microbes (ii) To prevent the spoilage of food (iii) To enhance the keeping quality of food (iv) To control food borne infections and intoxications (v)To extend the shelf life of food (vi) To reduce economic losses [12].

8. Examples of the hurdle effect

Every stable and safe food and food products are having several sets of intrinsic hurdles which differs in quality and intensity based on a particular product, however, in any case the hurdles must keep the 'normal' population of microorganisms in this food under control. The microorganisms present at the begging stage of food

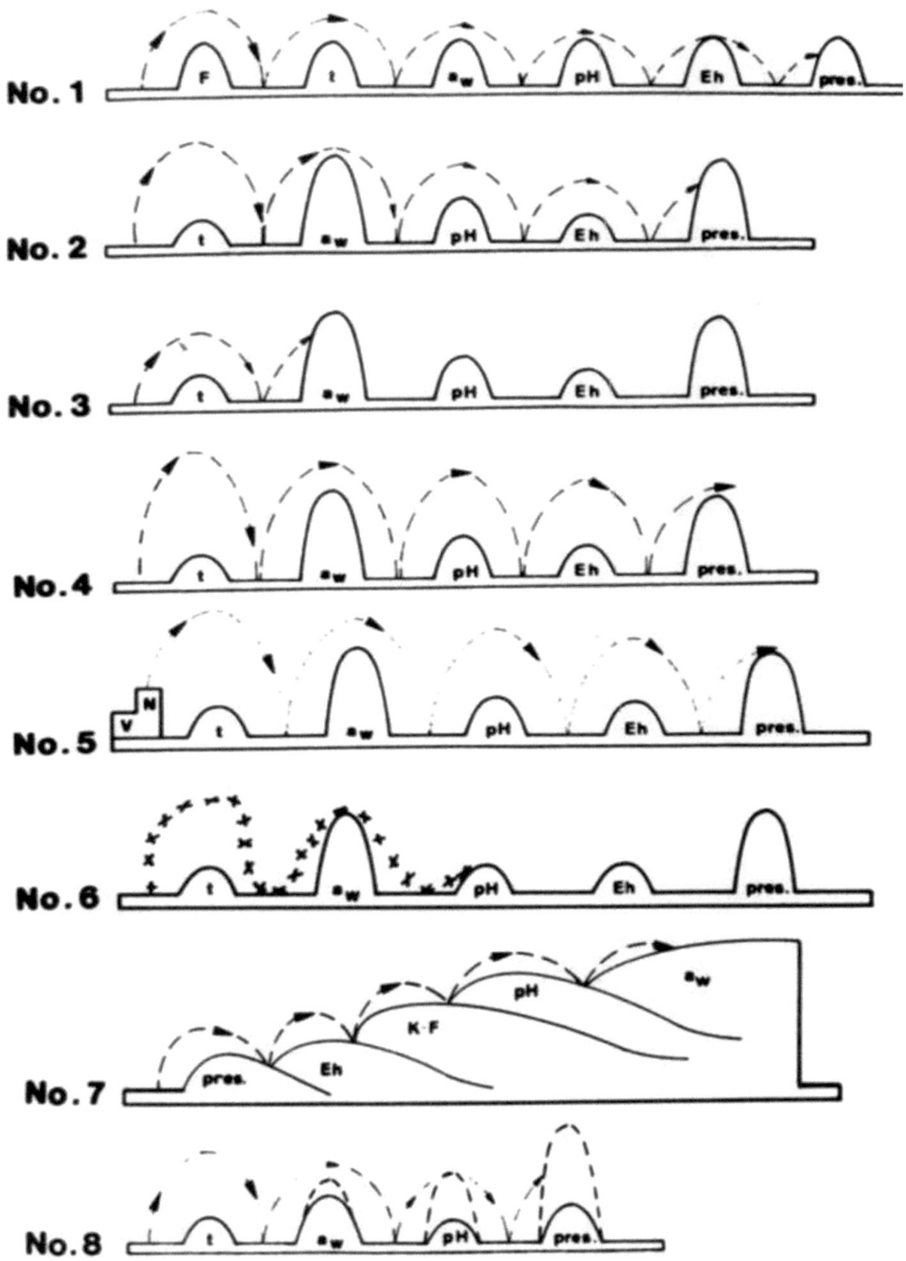

Figure 2.
Illustration of the hurdle effect, using eight examples. Symbols have the following meaning: F, heating; t, chilling; a_w, water activity; pH, acidification; eh, redox potential; pres., preservatives; K-F, competitive flora; V, vitamins; N, nutrients [7, 11].

products, are incapable to jump over the next hurdles present in the food systems. Few examples of the hurdle effects depicted in **Figure 2**.

8.1 Fish and fishery products preserved by hurdle technology

While applying hurdle technology for particular food product the selected processing method (preservation and packaging) will affect the spoilage mechanism and leads to the major quality deterioration fish and fishery product. For example, the fish preserved by cold smoking process the combination of hurdles includes salt addition, mild thermal treatment and storage at low temperatures, in this case the quality deterioration is mainly attributed to microbial spoilage, resulting in sensory modifications and thus organoleptic rejection. However, when same food products stored at subzero temperatures then the quality deterioration is correlated with physical and chemical reactions, viz., dehydration and lipid oxidation. These are major factors that defining duration of shelf life of the final products [13].

8.2 Fermented fish products

The Southeast Asian countries are well-known for fermented fish products and preservation, similarly in India the northeast region is famous for same [14]. These areas are in rich natural resources and a cauldron of different people and cultures lie deep in the lap of easternmost Himalayan hills [15]. A number of hurdles are used for preparation of the fermented fish products, these hurdles makes the food products more stable, with enhanced sensory quality and are safe at room temperature. According to Erkmen & Bozoglu (2016) [16], the storage stability of fermented fish products could be attained by applying a combination of hurdles at several stages of the manufacturing process.

8.3 Refrigerated fish products

The fish and fishery products preserved by freezing and stored under frozen condition, it can provide a storage life of more than one year, if appropriately carried out. It has enabled fishing vessels to remain at sea for long periods, and allowed the storing of fish during periods of good fishing and high catching rates, as well as extended the market for fish products of high quality (http://www.fao.org/3/v3630e/v3630e03.htm) [17]. However, it has been reported that the temperature conditions in the actual cold chain often deviate from the recommended range. For this reason, the preservation of refrigerated fish products is alone is not sufficient. During the past years, the fish mainly gutted fish and fillets are stored under modified atmospheres or in vacuum. According to Tsironi & Taoukis (2010), it has been found that the combined application of modified atmospheres with the low a_w by applying osmotic dehydration along with addition of nisin in the osmotic solution may significantly extend the shelf life of refrigerated gilt-head seabream fillets during storage at 0–15°C.

8.4 Thermally preserved fish products

Preservation by heat is a major method for extending the shelf life of packaged fish because of the advantages of a high safety level, convenience and a healthy product. In thermal processing, food is preserved in hermetically sealed containers in cooked form for storage at ambient temperature, without compromising on the quality. It's mainly depends on hurdle technology and the final products exhibit usually very long storage stability. According to Choulitoudi et al. (2017) [18] the combined application of hot smoking and edible coating based active packaging enhanced by the incorporation of rosemary essential oil and/or extract at refrigerated storage under vacuum on eel fillets.

9. Public health significance

The quality of fish and fishery products is of major concern to the seafood processors, consumers and public health authorities. The quality of fish degrades, due to a complex process mainly by physical, chemical and microbiological forms of deterioration are implicated. However, some sea foods are processed in a modern fish industry which is technologically advanced and complicated industry in line with any other sea foods industry and with the same risk of products being contaminated with pathogenic microorganisms [19]. However, the greatest risk to human health is due to the consumption of spoiled fish, improper processed, improper preserved fish and fish products. There are several methods of fish preservation and processing one of the among hurdle technology is applied in seafood processing which ensure seafood safety. This technology ultimately improves the public health from food posing, seafood borne pathogen, food borne illness.

10. Advantages of hurdle technology over other methods of preservation

- The main advantage of this technique is affinity to overcome the ability of microorganisms in developing resistance to conventional preservation methods since this technique using combination of different preservation technique which acts synergistically by hitting different targets within the cell of the spoilage microorganism.

- Basically, in this technique, hurdles are use at lower concentrations this prevent the undesirable sensory changes and also provide the lower production cost and save energy.

- Another advantage of this technique is using natural preservatives in combination with synthetic preservatives, this also lower the risk associated with using synthetic preservatives at high concentration.

- Possibility of increasing shelf-stable foods; because food preserved by combined methods (hurdles) remains stable and safe even without refrigeration, and is high in sensory and nutritive value due to the gentle process applied.

11. Limitations of hurdle technology

Hurdles used in seafood preservation can provide various degree of microbial stress reactions, these stress reactions or cross-tolerance may not work when multiple hurdles are used. Mainly three type of possible results while applying hurdles technology for seafood preservation [20].

i. Addition or additive effect,

ii. Synergism or synergistic effect,

iii. Antagonism or antagonistic effect.

Here the term additive effect imply that effects of the individual substances are simply added together. Generally, the combination of hurdles has a higher inhibitory effect than any single one. Synergistic effect means that the inhibitory action of the combination of hurdles at intensity lower than that of the constituent hurdles separately. In an antagonistic effect, the needed hurdle level is stronger than that of the single constituents. Sometimes combination treatments are not much effective in lowering microorganisms than single treatments. These effects of combined hurdles are antagonistic. In some cases, application of the hurdle technology for seafood preservation may inhibit outgrowth but induced tolerance capability of microorganism and hence extended their survivability [20].

12. Conclusion

This hurdle technology is an effective and simple method in food preservation fields but it requires strategic processes. It is an important approach that can be used to improve quality parameters during processing and storage of food. Hurdle technology Smart application of hurdles improve sensory characteristics, chemical and microbiological qualities of food. Hurdles in the food preservatives require varying results on the basis of microbial stress. Undoubtedly, it will help in fulfilling the demand for fresh and natural food products. It will actually slow the emergence of new routes of microorganisms in the food items and thus keeps the foods safe and healthy to eat.

Acknowledgements

All Authors are thankful to Dean College of Fisheries Science Gumla Birsa Agricultural University Kanke Ranchi Jharkhand for encouragement and guidance during writing of this book chapter.

Author details

Jag Pal*, Om Pravesh Kumar Ravi, Sangeeta Kumari and Akhilesh Kumar Singh
College of Fisheries Science Gumla, Birsa Agricultural University,
Ranchi, Jharkhand, India

*Address all correspondence to: jagpalfpt@gmail.com

References

[1] Robinson, R. K. (2014). Encyclopedia of food microbiology. Academic press.

[2] Leistner, L. (2000). Basic aspects of food preservation by hurdle technology. International Journal of Food Microbiology, 55, 181-186. Doi.10.1016/s0168-1605(00)00161-6.

[3] Leistner, L. (1985). Hurdle technology applied to meat products of the shelf stable product and intermediate moisture food types." Properties of water in foods. Springer, Dordrecht. 309-329.

[4] Leistner, L. (1995). Principles and applications of hurdle technology. In New methods of food preservation (pp. 1-21). Springer, Boston, MA.

[5] Aditya Pundhir and Nida Murtaza. (2015) Hurdle Technology-An Approach towards Food Preservation. Int. J. Curr. Microbiol. App. Sci 4(7):802-809

[6] Leistner, L. (2000). Basic aspects of food preservation by hurdle technology. International Journal of Food Microbiology, 55, 181-186. doi.org/10.1016/S0168-1605(00)00161-6

[7] Leistner, L. (1992). Food preservation by combined methods. Food Research International, 25, 151-158. doi.org/10.1016/0963-9969(92)90158-2

[8] Erkmen, O., & Bozoglu, T. F. (2016). Food microbiology: Principles into practice. Chapter 9: Food preservation by combination of techniques (hurdle technology) (1st ed.). John Wiley & Sons, Ltd.

[9] Tsironi, Theofania, Dimitra Houhoula, and Petros Taoukis. (2020). "Hurdle technology for fish preservation." Aquaculture and Fisheries doi.org/10.1016/j.aaf.2020.02.001

[10] Leistner, Lothar, and Leon GM Gorris. (1995) "Food preservation by hurdle technology." Trends in Food Science & Technology 6.2: 41-46. doi.org/10.1016/S0924-2244(00)88941-4

[11] Leistner, L. (1978) In Food Quality and Nutrition, ed Downey, W.K. Applied Science Publishers, London, p. 553. doi.org/10.1016/0963-9969(92)90158-2

[12] Tsironi, T., Houhoula, D., &Taoukis, P. (2020). Hurdle technology for fish preservation. Aquaculture and Fisheries. doi.org/10.1016/j.aaf.2020.02.001

[13] Majumdar, R. K., Roy, D., Bejjanki, S., & Bhaskar, N. (2016). An overview of some ethnic fermented fish products of the Eastern Himalayan region of India. Journal of Ethnic Foods, 3(4), 276-283. doi.org/10.1016/j.jef.2016.12.003

[14] Erkmen, O., &Bozoglu, T. F. (2016). Food microbiology: Principles into practice. Chapter 9: Food preservation by combination of techniques (hurdle technology) (1st ed.). John Wiley & Sons, Ltd. doi.org/10.1002/9781119237860.ch35

[15] Tsironi, T., Houhoula, D., &Taoukis, P. (2020). Hurdle technology for fish preservation. Aquaculture and Fisheries. doi.org/10.1016/j.aaf.2020.02.001

[16] http://www.dimorianreview.com/2016/11/hurdle-technology combined-method-for.html#ixzz6Rw5mRGXk

[17] Choulitoudi, E., Ganiari, S., Tsironi, T., Ntzimani, A., Tsimogiannis, D., Taoukis, P., &Oreopoulou, V. (2017). Edible coating enriched with rosemary extracts to enhance oxidative and microbial stability of smoked eel fillets.

Food packaging and shelf life, 12, 107-113. doi.org/10.1016/j.fpsl.2017.04.009

[18] Food and Agricultural Organization Technical Manual (2006). Handling and Preservation of Fruits and Vegetables by Combined Methods for Rural Areas. FOA Agricultural Service Bulletin 149. Pp 1,46

[19] Adedeji, O. B., Okerentugba, P. O., Innocent-Adiele, H. C., Okonko, I. O., Ojeniyi, S. O., Adejoro, S. A., & Mohamed, S. A. (2012). Benefits, public health hazards and risks associated with fish consumption. New York Science Journal, 5(9), 33-61.

[20] Erkmen, O., & Bozoglu, T. F. (2016). Food Microbiology, 2 Volume Set: Principles into Practice. John Wiley & Sons. doi.org/10.1002/9781119237860.ch35

Chapter 9

Dementia and Nutrition

Krishna Prasad Pathak and Emanuela Mattos

Abstract

Global aging population worldwide increasing. As growing age, the aging related issues like dementia came to be seen not as an inevitable condition at the old age phase, but as a condition that results from the competition between multiple risk factors and protective factors acquired throughout life. There is currently no cure for dementia. Thus, strategies to prevent or delay onset of dementia by changes in lifestyle factors, such as diet, are important as non-pharmacological therapy. A healthy nutrition contributes in delaying the cognitive decline for the elderly people and dementia patients. Cognitive decline is a normal part of the aging process and it is a main clinical identification in between elderly and dementia. The group of B Vitamins (B1, B2, B6, B12) are significantly associated with healthy neuropsychological function. The lack of B12 can show impairment of cognition and neurologic deficit and impacts on educational achievement. The cognitive impairment is a main clinical symptom of dementia which can raise the prevalence rates of cognitive impairment that can be dementia accordingly at the end of life.

Keywords: dementia, cognitive impairment, diet

1. Introduction

The aging of the population worldwide has accentuated the emergence of chronic degenerative diseases, including dementia. This is already recognized as a public health priority, given the social and economic impacts at local and regional levels. The absence of effective pharmacological treatments for the reversal of the conditions, caused many researches to be developed in search of preventive strategies. In the past few decades, researchers have focused on identifying potentially modifiable risk factors. This led to a paradigm shift. Dementia came to be seen not as an inevitable condition at the old age phase, but as a condition that results from the competition between multiple risk factors and protective factors acquired throughout life. Scholars have already pointed out evidence linking dementia to vascular risk factors (e.g., hypertension, diabetes, obesity), psychosocial factors (e.g., depression), lifestyle behaviors (e.g., low intellectual level and physical activity, smoking). Meanwhile, high level of education and work complexity, social networks, involvement in mental stimulating activities and regular physical exercise have shown protective properties against dementia [1–3]. Based on this, interventions focused on lifestyle modification have been conducted to assess the impact and association of eating habits and cognitive performance/dementia [1].

Data from the Alzheimer's International Disease (ADI) points out that the number of elderly citizens in the world comprises almost 900 million people. Most live in low-income countries. Longer life expectancy is one of the main risk factors for the development of chronic diseases, with dementia being the most prevalent in this population [4]. The figure bellow shows the high difference between the number of people with dementia in low and middle income compared to high income countries (**Figure 1**).

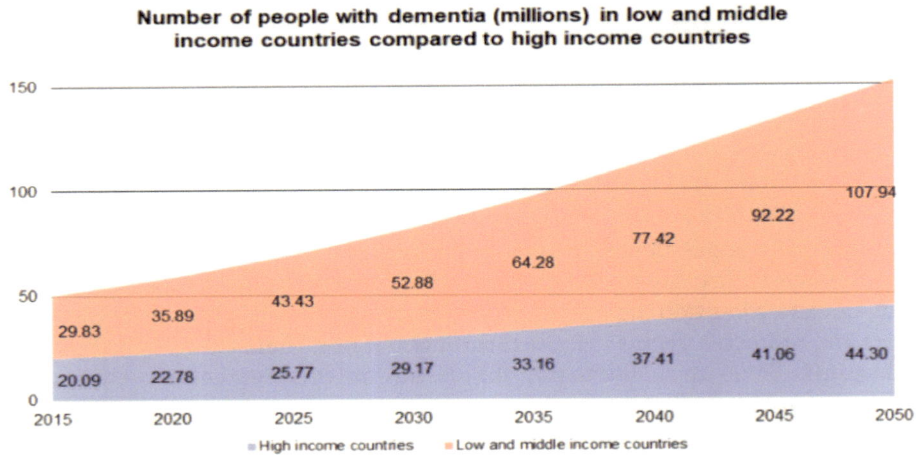

Figure 1.
Alzheimer's Disease International (2015) [5].

Aspects related to industrial and economic development and urbanization make traditional societies adapt to rapid changes. These 'social' aspects have been little discussed but their impacts as well as those caused by the demographic and epidemiological transition will make a profound difference in the great number of cases of dementia largely concentrated in middle and low-income countries. In 2015, the country with the highest percentage of elderly citizens was Japan (33.2%), and the lowest was Uganda (3.7%). Analysing the table below, it is possible to predict the distribution of the world's elderly population according to the country's income range in the coming years [4] (**Figure 2**).

The world's population of older people (age 60 and over, millions), and their distribution according to country income level (World Bank Classification 2009 and 2015)

	Current and projected numbers of older people (% of total population)					% increase over time	
Year	2010	2015		2030	2050	2015-2030	2015-2050
World Bank income classification	2009	2009	2015	2015	2015	2015	2015
HIC	232.3 (30.4%)	258.7 (28.9%)	309.4 (34.6%)	403.9 (29.4%)	482.5 (23.9%)	31%	56%
UMIC	116.4 (15.2%)	135.3 (15.1%)	319.8 (35.7%)	531.5 (38.7%)	760.8 (37.7%)	66%	138%
L-MIC	356.2 (46.6%)	431.7 (48.2%)	233.1 (26.0%)	386.0 (28.1%)	665.3 (32.9%)	66%	185%
LIC	59.8 (7.8%)	69.5 (7.8%)	32.9 (3.7%)	53.5 (3.9%)	111.4 (5.5%)	63%	239%
World	764.7 (100%)	895.2 (100%)		1347.8 (100%)	2020.0 (100%)	51%	126%

Figure 2.
Alzheimer's Disease International (2015) [4].

The biggest concern is that the poorest countries, of course, have less savings and professional human resources for the needs of health care and social assistance for populations that age rapidly and consequently for those who develop chronic diseases such as dementia [4]. Mayeda et al. (2016) pointed out the results of their study on inequalities in the incidence of dementia with six racial and ethnic groups over 14 years. The results showed that the dementia incidence was highest for African-Americans (26.6/1,000 person-years) and American Indian/Alaska Native (AIANs) (22.2/1,000 person-years); intermediate for Latinos (19.6/1,000 person years), Pacific Islanders (19.6/1,000 person-years), and Whites (19.3/1,000 person-years) and lowest among Asian-Americans (15.2/1,000 person-years). Risk was 65% greater for African Americans versus Asian-Americans. These inequalities in dementia incidence were observed among women and men and across all ages and estimated cumulative incidence of dementia over 25 years was high for all groups [6].

Worldwide, a significant portion of the elderly population has nutritional problems such as malnutrition (for example, micronutrient deficiencies, like B vitamins, vitamin C, E, D, Se, Zn, Ca and Fe) and over-nutrition (ie, obesity); often existing together [7]. A growing body of evidence has been focused on the association between dietary habits and cognitive performance/dementia. Numerous studies have pointed out that various dietary patterns and nutritional components such as (Mediterranean diet, unsaturated fatty acids, antioxidants (such as vitamin E, vitamin C and flavonoids, vitamin B) are associated with a significantly reduced risk of dementia [8]. In addition, low concentrations of vitamin D were related to an increased risk of cognitive decline [8]. The understanding that nutritional components can play a protective role and favorably influence the cognitive trajectory from the adoption of specific nutritional habits has been widely investigated. Although substantial progress has been made in identifying predictors of healthy aging the impact of certain interventions have proven to be increasingly challenging. This is due, in partly due to the cost and difficulty of studying the long latency disability and illness, and the subjective nature of some predictors of healthy aging.

Marsman et al. appointed: "Identify inequalities and identify social, economic and other policies that can reduce unjust processes within and between countries are essential to improving the lives of all chances of optimizing healthy aging. However, measuring healthy aging and the impact of interventions on life style remains an area of research focus. Metabolomics and transcriptomics led to the emergence of biomarkers that can be used to assess the impact of various interventions in the healthy aging process. Clearly, the role of diet and nutrition is central to maintaining life" [7].

2. Diet

There is currently no cure for dementia. Thus, strategies to prevent or delay onset of dementia by changes in lifestyle factors, such as diet, are important as non-pharmacological therapy. In, 2014, The World Alzheimer Report pointed the evidence from cross-sectional studies has shown that, compared to adults with dementia, healthy elderly people tend to have a healthier life diet, richer in fruits and vegetables, instead of meat, processed carbohydrates and fats [9]. It is not yet known why eating habits can contribute to the development of dementia, but studies that have been carried out so far have given rise to hypotheses fats [9].

A healthy nutrition contributes in delaying the cognitive decline for the elderly people and dementia patients [10]. Cognitive decline is a normal part of the aging process [11] and it is a main clinical identification in between elderly and dementia. The group of B Vitamins (B1, B2, B6, B12) are significantly associated with healthy neuropsychological function [12]. The lack of B12 can show impairment of cognition and neurologic deficit and impacts on educational achievements [13]. The cognitive impairment is a main clinical symptom of dementia which can raise the prevalence rates of cognitive impairment that can be dementia accordingly at the end of life.

ADI reinforces that dementia is a multifactorial chronic condition with a long latency period between the beginning of the complex pathophysiological and clinical mechanisms and the surgeries of the first symptoms, difficulty and limitation to the performance of definitive trials that test the effect of the diet on it [9]. Prospective cohort studies with long follow-up intervals can obtain associations in their results biologically plausible with epidemiological evidence that guide dietary recommendations to reduce the risk of dementia in populations [9]. Researchers recognize B vitamins as essential after cellular metabolism and that needs supplementary dietary intake because our bodies cannot synthesize necessary quantities. There are eight different chemically distinct types of vitamin B, with B6, B9 and B12 all being linked to protective roles in cognition [9], as well as exogenous antioxidants, including vitamins A (eg, b-carotene), C and E (tocopherols), and minerals, such as manganese, copper, selenium and zinc [14]. In 2004, Garcia and Zanibbi researched the mechanisms of action of homocysteine and speculated that the effects of elevated homocysteine on the brain may be irreversible, in cases where neuropathological changes were already observed [9, 15]. Studies identified that when folate or vitamin B12 are deficient, homocysteine levels rise, which may contribute to amyloid and tau protein accumulation and neuronal death. Homocysteine stimulates apoptosis and neurotoxicity (leading to nerve cell death), and platelet activation (contributing to white matter lesions, vascular injury and ischaemic strokes [9, 15]. The association between B vitamins and cognition has been the subject of several recent systematic reviews [9, 15]. Antioxidants has been the subject of investigation. Studies try to understand how they can inhibit the production of toxic substances and reduce the damage caused by free radicals and, consequently, neurodegeneration [9]. PUFA (polyunsaturated fatty acids) Omega-3 which are not synthesized by the human body, is an essential food constituent in view of its importance for the brain [9, 15]. Research suggests that Omega-3 may be involved in the vascular and inflammatory system and the amyloid pathways of dementia, and be therefore, potentially important in vascular dementia, Alzheimer's disease and mixed forms [9]. Evidence on the beneficial effects of fish consumption for preventing the incidence of dementia are inconsistent. However, healthy lifestyles and circumstances of life (including socioeconomic and educational level) that are associated with higher fish consumption and lower risk of dementia can explain the positive results found by some studies [9, 16]. The Mediterranean diet, which consists of a high intake of cereals, fruits, fish, legumes and vegetables, was associated with reduced risk for a number results, including cardiovascular disease, type 2 diabetes, some forms of cancer and in general mortality studies [9, 17]. The main biological mechanisms are related to the impact on the vascular system, oxidative stress and attenuation of the inflammatory pathway, proposals to support these associations can reduce the risk of cardiovascular disease, which in itself is a risk factor for dementia [9, 18–20].

The Mediterranean foods including olive Oil, consumption of fish (it is rich in omega-3), are important to maintain the proper brain functions [21] and regulate the oxidative stress [15, 22]. In Indian and Nepali food culture people use the daily cumin, Curcumin (Turmeric) in their kitchen which has been used since long century and is believed to heal various medical conditions like; gastric, ulcer, arthritis, liver disease including and traumatic brain injury and treatments for dementia [23, 24]. The another herbal called Gingo biloba prescribed to preserve the memory as medication for the dementia patients as the primary prevention [25]. It is also, suggest by vitro study that Gingo biloba helps as anti-amyloid aggregation effect and beneficial in dementia prevention [26]. A study finds that a higher intake of red meat in midlife was associated to increase the cognitive impairment at the end of life. However, the poultry/Sell-fish food reduced the risk in the population of Chinese people [27]. Likewise, lacking of iron is associated to be cognition impairment [10]. Animal study showed capacity of brain can be strong if the dietary of Iron is in balance. Similarly, a human model study documented that iron treatment helped positively to concentrate [11]. Iron deficiency in child age also risk to have poorer cognitive impairment in four countries study, India, Mexico, Zanzibar and Chile [12]. Besides this protein deficiency is also most important cause of poor attention, motivation, motor control and perception of the children that leads to the ends of life [13]. Thus, overall lacking of nutrients (B12, iron, protein, Zink, energy) can irreversible effects to the developing children and impacts cognitive functions and quality of life [28].

Inadequate childhood nutrition is associated with poor short-term academic and cognitive outcomes inadequate childhood nutrition is associated with poor short-term academic and cognitive impairment in later life, while substitution of red meat intake with poultry or fresh fish/shellfish was associated with reduced risk. While substitution of red meat intake with poultry or fresh fish/shellfish was associated with reduced risk [29].

3. Conclusion

Despite the association evidenced by the studies described so far between the protective role of nutrition and cognitive decline, no observational or randomized controlled study has been conclusive regarding the effectiveness of a type of nutritional intervention. As dementia is a multifactorial chronic condition, health professionals before the proportion of consumption of a given nutritional diet, should consider the nutritional resources accessible to the elderly according to social, economic, cultural, educational and regional aspects. While observational studies use self-report questionnaires that assess the individual at a single point in time and may have several variations in nutritional consumption that are not reported or that may have interference from experiences prior to evaluation, longitudinal studies (RCT) can adjust some variables such as the level of education of the subjects in their analysis, but indicators such as social and economic level are less frequently considered.

Author details

Krishna Prasad Pathak[1,2,3]* and Emanuela Mattos[3]

1 Nepal Open University, Nepal

2 Alzheimer and Related Dementia Society Nepal (ARDS-Nepal), Nepal

3 Federal University of São Paulo, São Paulo, Brazil

*Address all correspondence to: krishnapathak32@gmail.com

References

[1] Canevelli, M.; Lucchini, F.; Quarata, F.; Bruno, G.; Cesari, M. Nutrition and dementia: evidence for preventive approaches? Nutrients 2016, 8, 144; doi:10.3390/nu8030144

[2] Beydoun, M.A.; Beydoun, H.A.; Gamaldo, A.A.; Teel, A.; Zonderman, A.B.;Wang, Y. Epidemiologic studies of modifiable factors associated with cognition and dementia: Systematic review and meta-analysis. BMC Public Health. 2014, 14, 643.

[3] Norton, S.; Matthews, F.E.; Barnes, D.E.; Yaffe, K.; Brayne, C. Potential for primary prevention of Alzheimer's disease: An analysis of population-based data. Lancet Neurol. 2014, 13, 788-794

[4] Alzheimer's Disease International (ADI). World Alzheimer Report. The global impact of dementia: na analysis of prevalence, incidence, cost and trends. Introduction, chapter 1, p.6-9, 2015.

[5] Alzheimer's Disease International (ADI). World Alzheimer Report. The global impact of dementia: na analysis of prevalence, incidence, cost and trends. The global prevalence of dementia, chapter 2, p.10-29, 2015.

[6] Mayeda, ER.; Glymour, MM.; Quesenberry, CP.; Whitmer, RA. Inequalities in dementia incidence between six racial and ethnic groups over 14 years. Alzheimers Dement. 12(3):216-224, March.,2016.

[7] Marsman, D.; Belsky, DW.; Gregori, D.; Johnson, MA.; Dog, TL.; ·Meydani, S., Pigat, S.; Sadana, R.; Shao, A.; Griffiths, JC. Healthy ageing: the natural consequences of good nutrition—a conference report. European Journal of Nutrition (2018) 57 (Suppl 2): S15–S34. https://doi.org/10.1007/s00394-018-1723-0.

[8] Cao, L.; Tan, L.; Wang, H.-F.; Jiang, T.; Zhu, X.-C.; Lu, H.; Tan, M.S.; Tu, J.T. Dietary patterns and risk of dementia: A systematic review and meta-analysis of cohort studies. Mol. Neurobiol. 2015.

[9] Alzheimer's Disease International (ADI). World Alzheimer Report. Dementia and Risk Redution: an analysis of protective and modifiable factors. Lifestyle, chapter 4, p.42-65, 2014.

[10] Jiang YW, Sheng LT, Pan XF, Feng L, Yuan JM, Pan A, Koh WP. Meat consumption in midlife and risk of cognitive impairment in old age: the Singapore Chinese Health Study. European journal of nutrition. 2020 Jun;59(4):1729-1738.

[11] Bailey R.L., West K.P., Black R.E. The epidemiology of global micronutrient deficiencies. Ann. Nutr. Metab. 2015; 66:22-33. doi: 10.1159/000371618.

[12] Seshadri S., Gopaldas T. Impact of iron supplementation on cognitive functions in preschool and school-aged children-the indian experience. Am. J. Clin. Nutr. 1989;50:675-686. doi:10.1093/ajcn/50.3.675

[13] Piñero DJ, Li NQ, Connor JR, Beard JL. Variations in dietary iron alter brain iron metabolism in developing rats. J Nutr. 2000 Feb; 130(2):254-263.

[14] Garcia A, Zanibbi K. Homocysteine and cognitive function in elderly people. CMAJ 2004; 171(8): 897-904

[15] Coley, N.; Vaurs, C.; Andrieu, S. Nutrition and Cognition in Aging Adults. Clin Geriatr Med, 2015

[16] Larrieu S, Letenneur L, Helmer C, Dartigues JF, Barberger-Gateau P. Nutritional factors and risk of incident dementia in the PAQUID longitudinal

cohort. J Nutr Health Aging 2004; 8(3): 150-154

[17] Sofi F, Cesari F, Abbate R, Gensini GF, Casini A. Adherence to Mediterranean diet and health status: meta-analysis. Bmj 2008; 337: a1344

[18] Stampfer MJ. Cardiovascular disease and Alzheimer's disease: common links. Journal of internal medicine 2006; 260(3): 211-23. 62.

[19] Vitali C, Wellington CL, Calabresi L. HDL and cholesterol handling in the brain. Cardiovascular research 2014; 103(3): 405-413.

[20] Swanson D, Block R, Mousa SA. Omega-3 fatty acids EPA and DHA: health benefits throughout life. Advances in nutrition. 2012 Jan;3(1):1-7. doi: 10.3945/an.111.000893

[21] Polidori, M. C., Pratico, D., Mangialasche, F., Mariani, E., Aust, O., Anlasik, T., et al. (2009). High fruit and vegetable intake is positively correlated with antioxidant status and cognitive performance in healthy subjects. J. Alzheimers Dis. 17, 921-927. doi: 10.3233/JAD-2009-1114

[22] Otaegui-Arrazola, A., Amiano, P., Elbusto, A., Urdaneta, E., and Martinez-Lage, P. (2014). Diet, cognition, and Alzheimer's disease: food for thought. Eur. J. Nutr. 53, 1-23. doi: 10.1007/s00394-013-0561-3

[23] Healton EB, Savage DG, Brust JC, Garrett TJ, Lindenbaum J. Neurologic aspects of cobalamin deficiency. Medicine. 1991 Jul 1;70(4):229-245.

[24] Heys M, Jiang C, Schooling CM, Zhang W, Cheng KK, Lam TH, Leung GM. Is childhood meat eating associated with better later adulthood cognition in a developing population?. European journal of epidemiology. 2010 Jul 1;25(7):507-516.

[25] Mishra S, Palanivelu K. The effect of curcumin (turmeric) on Alzheimer's disease: An overview.Annals of Indian Academy of Neurology. 2008 Jan;11(1):13.

[26] DeKosky ST, Williamson JD, Fitzpatrick AL, Kronmal RA, Ives DG, Saxton JA, Lopez OL, Burke G, Carlson MC, Fried LP, Kuller LH. Ginkgo biloba for prevention of dementia: a randomized controlled trial. Jama. 2008 Nov 19;300(19):2253-2262.

[27] Luo Y, Smith JV, Paramasivam V, et al. Inhibition of amyloid-beta aggregation and caspase-3 activation by the Ginkgo biloba extract EGb761. Proc Natl Acad Sci U SA. 2002;99(19): 12197-12202.

[28] Kar BR, Rao SL, Chandramouli BA. Cognitive development in children with chronic protein energy malnutrition. Behav Brain Funct. 2008 Jul 24; 4():31.

[29] Sharifi-Rad J, Rayess YE, Rizk AA, Sadaka C, Zgheib R, Zam W, Sestito S, Rapposelli S, Neffe-Skocińska K, Zielińska D, Salehi B. Turmeric and its major compound curcumin on health: bioactive effects and safety profiles for food, pharmaceutical, biotechnological and medicinal applications. Front Pharmacol. 2020; 11: 01021. doi: 10.3389/fphar.2020.01021.

Chapter 10

Meat Borne Diseases

Dhary Alewy Almashhadany

Abstract

Red and white meat is a perfect, high-quality protein that comprises all of the nine essential amino acids (EAAs) that cannot be synthesized endogenously. Meat is the normal source of this vitamin, as well as other types of vitamins. Meat affords a range of significant vitamins and minerals that the human body needs, many of which are more bioavailable and easily absorbed than the nutrients found in plant sources. The nutrients in meat support the immune system, participate in the formation of muscle tissue, red blood cells (RBCs), and hormones, and warranty accurate functioning of the nervous system. These nutrients also affect the human senses of smell and taste, benefit our thyroids, and support antioxidant production. The main sources of pathogens in meat and meat products are; the animal itself; human handlers; equipment's in contact, environmental sources, and water used in the preparation. Meat Borne Diseases, since ancient times, played a central role in public health. This chapter is divided into nine parts, part one to part eight deals with the most important pathogens that have been associated with meat borne diseases (MBDs), these include, Meat Borne Prionic Diseases; Viral Diseases; Bacterial Diseases; Protozoal Diseases; Parasitic Diseases; Fungal Diseases; Mycotoxins; Rickettsial Diseases; while the nine-part deal with the methods of meat preservation and storage.

Keywords: meat, diseases, preservation

1. Introduction

Meat is the flesh of certain animal species that are used as food by humans. The main tissue is the skeletal muscle and its associated tissues, also the edible offal which includes organs and non-skeletal muscle tissues are considered meat. It is derived from avian, mammalian, reptilian, amphibian, and aquatic species commonly harvested for human consumption [1, 2].

Meat can be generally categorized as red or white depending on the concentration of myoglobin in muscle fiber. Meat is mainly composed of water, protein, and fats, followed by minerals, vitamins, carbohydrates, and other bioactive components [3]. From the nutritional point of view, meat's significance is derived from its high-quality protein, containing all essential amino acids, and its highly bioavailable minerals and vitamins. Meat is affluent in Vitamin B12 and iron which are not easily available in botanical foods [4]. Meat has played a vital role in human development and is a vital constituent of a well-balanced diet. It ranks among one of the most important, nourishing, and preferred food items available to the masses, which aids in fulfilling most of their body necessities. Meat fat and its fatty acid profile is a point to worry, concerning its consumption, but its moderate custom is always advised by physicians and nutritionists, to lead a healthy life. The fat content of animal carcasses ranges between 8 and 20%. The average proportion of meat protein is about 23% that fluctuates from a lower to a higher value according to the type of meat source [5]. Generally Meat is a perfect, high-quality protein comprise all of the nine essential amino acids (EAAs) that cannot be synthesized endogenously (lysine, isoleucine, methionine, leucine, valine, tryptophan, threonine, phenylalanine, and histidine) cannot be synthesized endogenously, the human body needs to consume to build and rebuild every cell in the body, as well as for optimal health [6]. Meats Vitamin B12 plays a key role in normal metabolism, preserving brain and nervous system function, and high energy levels. Meat is the normal source of this vitamin, as well as other types of vitamins.. Meat affords a range of significant vitamins and minerals that the human body needs, many of which are more bioavailable and easily absorbed than the nutrients found in plant sources. The nutrients in meat support the immune system, participate in the formation of muscle tissue, red blood cells (RBCs), and hormones, and warranty accurate functioning of the nervous system. These nutrients also affect the human senses of smell and taste, benefit our thyroids, and support antioxidant production [7, 8]. Muscular tissue in live healthy animals is virtually free any contaminant microorganisms. However, following animal slaughtering and carcass preparation, muscular tissue is being subjected to various microorganisms [9, 10].

Depending on different sanitary conditions prevailed upon meat preparation meat might be subjected to different pathogens which might be transmitted to humans (Meat Borne Diseases) [11]. These include disease causative agent like; Prion, Viral, Bacterial, Mycotic, and Parasitic Diseases [12]. The main sources of pathogens in meat and meat products are; the animal itself; human handlers; equipment's in contact, environmental sources or water used in preparation [13]. Therefore, strict hygienic precautions must me prevailed during meat handling and preparations. Meat and its products have been engaged in many diseases or outbreaks in human consumers which necessity awareness and educational knowledge about causative agents and control hygienic measures. This chapter will cover the most important pathogens that have been associated with meat borne diseases (MBDs).

2. Meat borne prionic diseases

These are a group of diseases caused by Prion, which are very significant in the field of public health, whether human public health or veterinary public health, that is commonly known as group of diseases Spongiform Encephalopathy [14]. The most important prionic disease transmitted from cattle to human through cattle meat is the Bovine Spongiform Encephalopathy (BSE) (Mad - Cow Disease) [15]. The diseases that was discovered for the first time in Britain in November of 1986, and it had infected cows, sheep, cats, and monkeys [16].

2.1 The causative agent

The incubation period is usually very long, ranging between (2–8) years [17]. Prion in infected cattle were found in brain tissues, and in the spinal cord, bone marrow, spleen, lymph nodes, tonsils, in addition to the intestine. Prion is infectious proteins that were previously called slow viruses (Slow Viruses), but they are similar to a virus in that they contain a protein and live and multiply inside the cell, taking into account that prions differ from viruses in that they do not have DNA in their composition or it may exist, but in small quantities. Prion, which causes mad cow disease, is characterized by It's a superior ability to resist heat, disinfectants, and UV rays and high ability to resist freezing, drying, and cooking temperatures [18].

2.1.1 Occurrence of diseases

It occurs as a result of cows eating diets containing animal protein remains, including meat and bone meal, which carry the pathogens. The occurrence of the disease started since 1970, where expansion began in Britain by using the carcasses of sick and dead animals to produce feed additives such as meat and bone meal. The preventive measures that have to be taken to facing the transmission of the disease to humans is to excluding and burning all animals that are proven to have the disease beside forbidding the use of mammalian meat and bones in feeding farm animals. At the same time all necessary health measures should be taken in red meat slaughterhouses, and emphasize the removal of animal waste and other wastes immediately after completion of the slaughtering and processing [19].

3. Meat borne viral diseases

Several viruses can cause foodborne illness, including meat and meat products. The most significant viruses transmitted to humans via foods comprise noroviruses, rotaviruses, adenoviruses, sap viruses, and astroviruses [20].

3.1 Hepatitis

Hepatitis A is caused by an infection with the hepatitis A virus (HAV). This type is most commonly transmitted by consuming water or food including meat and meat products contaminated by feces from a person infected with hepatitis A [21].

Hepatitis B caused by an infection with the hepatitis B virus (HBV). This type is transmitted through contact with infectious body fluids, such as blood, semen, and vaginal secretions, containing the (HBV). Injection drug use, having sex with an infected partner or sharing razors with an infected person increase the risk of getting hepatitis B [22].

Hepatitis C is caused by an infection with the hepatitis C virus (HCV). This type is transmitted through direct contact with infected body fluids, typically through injection drug use and sexual contact [23]. HCV is among the most common blood-borne viral infections in some countries, like USA.

Hepatitis D (delta hepatitis) It is an infection with the hepatitis D virus (HDV). HDV is contracted through direct contact with infected blood [24]. The hepatitis D virus cannot multiply without the presence of hepatitis B. It is a rare form of hepatitis that only occurs in conjunction with hepatitis B infection.

Hepatitis E caused by infection through the hepatitis E virus (HEV). Hepatitis E is mainly found in areas with poor sanitation and typically results from ingesting fecal matter that contaminates the water supply [25]. Cases of hepatitis E have been indicated in the Middle East, Asia, Africa, and Central America.

3.1.1 Noroviruses

The infection occurs through oral ingestion from contaminated food including meat and meat products, as well as water. The transmission also occurs through aerosols creating during vomiting and fomites. However, the primary route of transmission is person-to-person transmission through the fecal–oral and vomit-oral routes, and indirectly through food (ready to eat including leafy vegetables and herbs, berries, and foods handled after cooking), water, and the environment.

The European Union summary report on trends and sources of zoonoses, zoonotic agents, and food-borne outbreaks in 2016, mentioned that food is implicated in up to 24% of global outbreaks. Crustaceans, shellfish, mollusks, and their products beside vegetables and juices are the foods most often implicated in European norovirus outbreaks in 2016 [26].

3.1.2 Rotaviruses

The virus affects mainly infants and young children. Rotaviruses cause enteric disease with symptoms characterized by fever, vomiting, diarrhea, and abdominal discomfort [27].

3.1.3 Group a rotaviruses

Group A rotaviruses are the most important agents of severe diarrhea in infants and young children and are prevalent worldwide. It is the major pathogens in humans and animals. Ten serotypes of human group A rotaviruses are defined by neutralization of one (VP7) of the two outer capsid proteins [28].

3.1.4 The non-group a virus

The non-group A viruses are divided into groups B, C, D, E, F, and G based on distinct group antigens. Of the non-group A rotavirus, only groups B and C have been detected in humans; they are not an important cause of disease in infants and young children [29].

4. Meat borne bacterial diseases

Because of the great health risks Red meats and white meat come from warm-blooded animals and, as such, their microbial flora is heterogeneous, consisting of mesophilic and psychrotrophic bacteria. These bacteria include pathogenic species from the animal and birds themselves, as well as from the environment, together with bacterial species introduced during slaughter and processing of raw products [9]. Most of these diseases are zoonotic diseases, which are transmitted to human beings, either directly or indirectly, and hence the meat and its products play an important role in transmitting these pathogens. Meat borne diseases are classified into meat borne infection and meat borne intoxication [30].

4.1 Meat borne infections

Meat borne infections are caused by the entrance of pathogenic bacteria contaminating meat and meat products into the body, and the reaction of the body tissues to their presence [12]. Meat borne infections tend to have long incubation periods and are usually characterized by fever. Bacterial meat borne infections include the following important pathogens.

4.1.1 Campylobacteriosis

The incubation period ranges between 2 and 11 days with an average of 3–5 days. *C. jejuni and C. coli* causes illness characterized by fever; abdominal pain (abdominal pain is associated with backache and possible mortality); foul-smelling and watery diarrhea, which runs for 3–4 days, (diarrhea may sometimes contain blood and mucus in feces); vomiting; nausea; and abdominal complaints [31].

4.1.2 Escherichia coli *foodborne infection*

E. coli is a common member of the normal flora of the large intestine [32]. Six pathotypes of *E. coli* are now recognized.

4.1.2.1 Enterohemorrhagic E coli *(EHEC)*

Enterohemorrhagic *E coli* (EHEC) causes haemorrhagic colitis or haemolytic-uremic syndrome (HUS) [33].

4.1.2.2 Enteroinvasive E coli *(EIEC) causes a Shigella-like dysentery*

EIEC strains cause illness that is characterized by watery diarrhea in most patients. Besides, there is a fever, nausea, and abdominal cramps [34].

4.1.2.3 Enteropathogenic E coli *(EPEC)*

Enteropathogenic *E coli* (EPEC) is a cause of childhood diarrhea [35]. The World Health Organization (WHO) estimated that every year 600 million (almost 1 in 10 people) fall in sick and nearly 420,000 deaths occurs worldwide as a result of contaminated food consumption [36].

4.1.2.4 Enterotoxigenic E coli *(ETEC)*

Enterotoxigenic *E coli* (ETEC) is a cause of traveler's diarrhea. ETEC are a pathogenic variant or pathovar of *E. coli* defined by production of diarrheagenic heat-labile (LT) and heat-stable (ST) enterotoxins [37].

4.1.2.5 Enteroaggregative E coli *(EAgg EC)*

Enteroaggregative *E coli* (EAgg EC) is primarily associated with persistent diarrhea in children in developing countries. Polluted food appears to be the main source of EAEC infection and has been associated in numerous foodborne outbreaks of diarrhea [38].

4.1.2.6 Enteroadherent E coli *(EAEC)*

Enteroadherent *E coli* (EAEC) is a cause of childhood diarrhea and traveler's diarrhea in Mexico and North Africa.

4.1.3 Listeria monocytogenes

Listeria monocytogenes is the only known species in the *Listeria* genus that concern for human health. It is G+ ive bacteria, that is pathogenic to both humans and animals [39, 40].

4.1.4 Salmonella species

Some of the important salmonella species involved in food poisoning include; *S. typhimurium, S. infantis, S. dublin, S. enteritidis, S. softenburg, S. montevideo, S. virchow, and S. Newport* [41]. Factors associated with Salmonella meat poisoning outbreaks include; consumption of inadequately cooked or thawed meat or poultry, cross-contamination of meat and meat products from infected food handlers besides the possible presence of rats, cockroaches, flies, in the food environment that acts as vectors of the disease [42, 43].

4.1.5 Shigellosis (bacillary dysentery)

All strains of shigella possess potent exotoxins which are carbohydrate-lipid protein complexes [44]. Any type of food including meat and meat products can transmit the shigella pathogens to cause disease in humans. Flies can spread Shigella germs when they get into contact with infected stool and then contaminate different types of food and drinking water. The illness begins 1 to 4 days after ingestion of bacteria and may last 4 to 7 days [45].

4.1.6 Vibriosis

4.1.6.1 Vibrio parahemolyticus

V. parahemolyticus is a pathogenic bacterium, whose natural habitat is the sea. Human infections occur solely from seafood creatures such as oysters, shrimps, crabs, lobsters, clams, and related shellfish.

4.1.6.2 Vibrio vulnificus

V. vulnificus causes severe foodborne infection. *V. vulnificus* infections can cause fever, nausea, myalgia, and abdominal cramps, 24–48 hours after eating contaminated food.

4.1.6.3 Vibrio cholera

Cholera is an infection of crowded poor class communities and it tends to persist in such areas. Human is the only natural host of the cholera [46]. The spread of infection is from person-to-person, through contaminated water or foods. Shrimps and vegetables are the most frequent carriers. Cholera is typically categorized by the sudden onset of uncomplicated vomiting, which is seen frequently, but very rapid dehydration and hypovolemic shock, as well as copious watery diarrhea. The frequent watery stools may be accompanied by small parts of the mucosa being liberated from the intestines [47].

4.1.7 Yersinia enterocolitica

Y. enterocolitica has been isolated from different types of food, such as beef, lamb, seafood, pork, milk, vegetables, and vacuum-packed meat [48]. Symptoms develop some days following ingestion of contaminated foods. It includes headache, fever, abdominal pain, diarrhea, and pharyngitis. Children appear to be more susceptible than adults [49].

4.1.8 Brucellosis

Brucellosis is a foodborne and professional zoonotic disease, caused by the bacterial genus Brucella. This infection has an extremely emerging and significant reemerging potentials in several countries [50].

Brucellosis is a cosmopolitan bacterial zoonotic disease (caused by Brucella spp.) that affects humans and various species of the wild and domestic animals, principally food-producing animals, including cattle, buffaloes, camels, sheep, goats, pigs, and reindeer [51, 52].

Human brucellosis is a severely debilitating and disabling life-threatening disease. It is recognized by the clinical problems such as, the contribution of the interior organs, peripheral arthritis, bronchopneumonia, epididymitis, orchitis, hepatic abscesses, sacroiliitis, osteomyelitis, spondylitis, meningitis, encephalitis, cardiovascular complications, and prostatitis [53].

The transmission occurs through ingestion of polluted milk or meat and from mothers to breastfed babies. The transmission of Brucella also happens through mucous membranes or skin wounds, following direct contact with urine, vaginal discharges, blood, tissues, placenta, aborted fetuses, and through inhalation of airborne agents in an atmosphere [54].

4.2 Meat borne intoxications

These are diseases caused by the consumption of meat, meat products, and other types of foods containing the following toxicants.

4.2.1 Bio toxicants

Poisonous animals and Plants, which are found in tissues of certain animals and plants [55].

4.2.1.1 Toxic fishes

Types of intoxications associated with fish include ciguatera poisoning, tetraodon poisoning, and scombroid toxicity [56]. They include puffers, triggerfish, and parrotfish.

4.2.1.2 Mollusca

Mollusca involved are oysters, mussels, and clams, which feed on dinoflagellates and planktons containing alkaloids making them toxic.

4.2.1.3 Poisonous mammals

Mammals are not commonly inherently poisonous, but secondary toxicity may affect many of them. The toxin may be of various types e.g. heavy metals, pesticides, toxic plants, therapeutics, fungal or bacterial toxins. Most human poisoning involves secondarily transfected toxins [57].

4.2.2 Metabolic products (toxins)

Metabolic products (toxins), which formed and excreted by microbes (Bacteria, Fungi, and Algae), while they multiply in the gastrointestinal tract (GIT) of human or in food [58].

4.2.3 Poisonous substances

Poisonous substances, which may be purposely or accidentally added to food during processing, production, transportation, or storage [59]. In general, the foodborne intoxications have short incubation Periods, from minutes to hours, and are characterized by a lack of fever. Food-borne intoxications can be classified into the following categories; Bacterial, Fungal and Chemical intoxications.

This is a type of meat-borne intoxication arising from ingestion of meat, meat products, and other types of food containing poisonous chemicals, such as heavy metals; Pesticides; insecticides; Herbicides; Fungicides. Chemicals also include Preservatives (Nitrites; antibiotics - penicillin, tetracycline, and chloramphenicol or Radionuclides (cesium, strontium, radium, barium, lanthanum) [60–62].

4.3 Bacterial meat borne intoxications

4.3.1 Bacillus cereus meat borne intoxication

Food poisoning caused by *B. cereus* is an acute intoxication that occurs when this bacterium produces toxins [63]. *B. cereus* is considered a comparatively common cause of gastroenteritis globally. There are two types of gastrointestinal disorders caused by this bacterium [64].

4.3.1.1 Emetic toxin (ETE)

The emetic syndrome, due to ETE, is an intoxication that is caused by a single highly heat-, proteolysis-, acid- and alkali-resistant toxin, that is pre-formed when ingested, leading to a rapid onset of the syndrome [65].

4.3.1.2 Hemolysin BL (Hbl)

Bacillus cereus produces emetic toxin and several enterotoxins including non-hemolytic enterotoxin (Nhe), hemolysin BL (Hbl), cytolysin K (CytK), hemolysin II (HlyII), enterotoxin FM (EntFM), and enterotoxin T (bc-D-ENT) .

4.3.1.3 Non-haemolytic enterotoxin (Nhe)

Is a pore forming toxin consisting of two lytic elements NheA and NheB, and a protein NheC with unknown function encoded by nheA, nheB, and nheC, respectively [66].

4.3.1.4 Cytotoxin K (CytK)

4.3.1.4.1 Emetic disorder

Characterized by vomiting, abdominal cramps, nausea, and occasionally diarrhea that occur 1–6 hrs after consumption of contaminated meat or other types of food.

4.3.1.4.2 Diarrhea disorder

Characterized by abdominal cramps, watery stool (copious diarrhea), tenesmus rarely vomiting. These symptoms beginning 8 to 16 hrs after ingestion of contaminated food.

4.3.2 Clostridium perfringens meat borne intoxication

Clostridium meat borne intoxication is caused by the ingestion of food containing large numbers of vegetative cells of enterotoxigenic *C. perfringens* type A and some type C and D strains. *C. perfringens* multiply in the intestine and sporulate releasing *C. perfringens* enterotoxin (CPE).

4.3.3 Clostridium botulinum *meat borne intoxication*

The danger of botulism has been the deciding factor in the formulation of food processing techniques, especially canned meat [67].

4.3.4 Staphylococcus aureus meat borne intoxication

Caused by the consumption of food including meat and meat products polluted with staphylococcal enterotoxins produced by confident strains of *Staph. aureus* while growing in different types of food [68]. These enterotoxins are pH stable (insensitive to pH changes); as well as resistant to most proteolysis enzymes (pepsin, renin, trypsin, and chymotrypsin) [69].

5. Meat borne protozoal diseases

5.1 *Toxoplasma gondii*

Human toxoplasmosis occurs from eating inadequately cooked meat, particularly mutton (lamb meat), pork, and venison (deer meat), or from drinking unpasteurized milk contaminated with *Toxoplasma gondii*. However, cooking meat (internal temperature about 70°C or 160°F) or freezing to around (−18°C or 0°F) should be able to destroy the protozoa [70]. Eating food that was contaminated by knives, utensils, cutting boards, or other foods that had contact with raw, contaminated meat or shellfish is other possible way too [71].

5.2 Sarcocystosis

Humans become infected when they eat undercooked meat comprising these Sarcocystis. Bradyzoites are released from ruptured cysts in the small intestine and enter the lamina propria of the intestinal epithelium. There, they distinguish into macro- and microgametocytes. The Union of male and female gametes results in the creation of oocysts. Oocysts sporulate in the intestinal epithelium and are shed from the host in feces [72].

5.3 Cryptosporidiosis

This protozoal diarrheal disease caused by *Cryptosporidium.* Both the protozoa and the disease are generally known as "Crypto." [73].

The probable hazards from meat borne cryptosporidiosis come from ingesting raw and uncooked foods, particularly meat and meat products. The foodborne transmission has been stated following the consumption of certain foods, such as uncooked meat products, raw sausage, offal, chicken salad, and milk. as well as the significance of disease confirmed by some researchers [74–76].

6. Meat borne parasitic diseases

6.1 Taeniasis

Eating raw or undercooked contaminated beef or pork is the primary risk factor for acquiring taeniasis. So, one way to prevent taeniasis is to cook meat at safe temperatures [77].

6.2 Trichinellosis/trichinosis

It occurs when a human eating raw or undercooked meat from animals infected with the protozoa Trichinella. Meat that comprises infective *Trichinella* larvae; the acid in the stomach dissolves the hard covering of the cyst around the larvae and releases the worms [78].

6.3 Diphyllobothriasis

Humans got infections by eating raw or undercooked fish. Examples of fish include salmon, trout, perch, walleyed pike, and other species of freshwater fish. Some fish such as salmon live in both fresh and saltwater and can harbor *Diphyllobothrium* larvae. Lightly salted, smoked, or pickled fish also may contain infectious organisms [79]. However, Cooking fish sufficiently, to an internal temperature of at least 145° F [~63° C]; or freezing at −4°F (−20°C) or below for 7 days (total time); or at −31°F (−35°C) or below until solid, and storing at −31°F (−35°C) or below for 15 hours; or at −31°F (−35°C) or below until solid and storing at −4°F (−20°C) or below for 24 hours [13].

6.4 Anisakiasis

Anisakiasis, or herring worm disease, is a parasitic disease caused by nematodes (worms) that attach to the wall of the esophagus, stomach, or intestine. Humans are accidentally infected when hosts are consumed either as raw or inadequately cooked or treated fish/shellfish meals. Therefore, the infection has been directly linked to eating habits [80].

6.5 Capillariasis

When humans ingest raw or undercooked infected fish, larvae may migrate to the intestine and mature into adult worms [13, 81].

6.6 Opisthorchiasis

Liver flukes infect the liver, gallbladder, and bile duct in humans. While most infected persons do not show any symptoms, infections that last a long time can result in acute symptoms and critical disease. Chronic infection may lead to cholangiocarcinoma, a cancer of the bile ducts [13, 82].

6.7 Heterophyiasis

Heterophyiasis is caused by trematode parasites happening in regions where brackish water fish is ingested raw or under inadequately cooked circumstances [83].

6.8 Clinostomiasis (yellow grub disease)

This parasite has a complex life cycle, usually taking mollusks and fishes as intermediate (middle) hosts and birds as final (definitive) hosts. Humans may become the definitive host by ingesting raw or undercooked fish meat infected with the metacercarial stage of this type of parasite [84, 85].

7. Meat borne fungal diseases

Fungi are very common in food because it being ubiquitous. It can spoil large amounts of food and produce hazardous toxins that threaten human health. However, yeasts and mold can grow in a large diversity of food including meat and meat products, which provide a favorable place for their growth [86]. The most significant pathogenic fungi have been isolated from a wide range of foods include the following.

7.1 Aspergilli

Aspergillus contains some species with strains that are the most dangeHP, with Aspergillus fumigatus causing the most serious diseases [87].

7.2 Fusarium

Well-known Fusarium mycotoxins are fumonisins, deoxynivalenol, zearalenone, and trichothecenes [88].

7.3 Mucor

Mucor contaminated food constitutes a limited potential health hazard concerning healthy consumers. No specific mycotoxin has been isolated and characterized in Mucor. The results of bioassays did indicate that toxins are present in extracts from certain Mucor species [89].

8. Mycotoxins

8.1 Aflatoxins (AFs)

The name AFs has been subsequent from the combination of "A" for the *Aspergillus* genus and "f" for the species *flavus*. AFs are greatly toxic, teratogenic, mutagenic, and carcinogenic compounds, produced as secondary metabolites by fungi belonging to numerous *Aspergillus* species, chiefly *A. flavus* and *A. parasiticus.*

Presently, 20 diverse categories of AFs have been recognized, wherein the main ones comprise AFB1, B2, G1, G2, and M1. Fungal species belonging to *A. flavus* naturally produce AFB1 and AFB2, while *A. parasiticus* can produce AFG1 and AFG2 in addition to AFB1 and AFB2 [90].

8.2 Fumonisins

It is the secondary metabolites of the *Fusarium* fungi mostly from *Fusarium verticillioides* and *Fusarium proliferatum* on pollute maize and milled maize portions or other processed products [88].

8.3 Ochratoxin a (OTA)

It is produced by *Penicillium verrucosum* in moderate environments and *Aspergillus ochraceus* and the rare *Aspergillus carbonarius* in warm and tropical countries that can pollute crops previous to yield and or more normally through storing [91].

8.4 Patulin (PAT)

Created by fungal species of the genera, *Aspergillus, Penicillium, and Byssochlamys, and the* most significant Patulin (PAT) producer is *Penicillium* [92].

8.5 Zearalenone (ZEA)

Non-steroidal estrogenic mycotoxin formed by a diversity of *Fusarium* fungi in comfortable and warm countries. ZEA presents a similar structure to estrogen and therefore competes with 17-estradiol in binding to the estrogen receptor. So, it can cause important differences in generative structures and fertility loss in humans and animals [93].

9. Meat borne rickettsial diseases

9.1 Query fever (Q-fever)

Persons get sick by inhalation dust that has been polluted by infected animal milk, urine, feces, and birth products that contain *Coxiella burnetii*. Individuals may get ill with Q fever by consuming contaminated, unpasteurized milk, and dairy products. Infrequently, Q fever has been spread through blood transfusion, from a pregnant woman to her foetus, or through sex [43, 94].

10. Meat preservation and storage

Meat preservation helps to control spoilage by hindering the growth of microbes, it delays spoilage; also reducing enzymatic activity, and avoiding the oxidation of fatty acids that stimulate rancidity, resulting in extends the life of the product; improves product quality. Several factors are affecting the period of meat storage. The physical state of meat acting a role in the number of microbes that can grow on meat, for example, grinding meat increases the surface capacity, releases moisture and nutrients from the muscle fibers, and distributes exterior germs throughout the meat. Chemical properties of meat, such as pH and moisture content, affect the capability of microbes to grow on meat. Usual protecting tissues, such as skin or fat, can prevent microbial pollution, dryness, or other disadvantageous fluctuations. Wrapping meats with paper or protecting plastic films avoids unnecessary moisture loss and microbial pollution. There are several methods for meat preservations [95].

10.1 Chilling/refrigeration

Temperature is the most significant factor in manipulating bacterial growth. Pathogenic bacteria do not grow well in temperatures under 3°C (38°F). So, meat should be stored at temperatures that are as cold as possible. Chilled packing is the most public method of meat preservation. The typical chilled packing life for fresh meats is 5 to 7 days.

10.2 Freezing

Freezer storage is an excellent technique of meat preservation. It is significant to covering frozen meats carefully in wrapping that limits air contact with the meat to avoid moisture loss during packing.

10.3 Meat curing

The commonly used technique of preserving meat before the days of chilling. It is done for communicating specific color and flavor development, as well as the preservative outcome. The main constituents comprise common salt (sodium chloride), sodium nitrate, sodium nitrite, and sugar.

10.4 Meat smoking

Smoking and curing of meat are consistent. Smoke generation is accompanied by the creation of several organic compounds (aldehydes, ketones, organic acids, phenols, etc.) and their concentration products. Phenols act as bacteriostatic; formaldehyde as a bactericidal compound, also informing typical smoky flavor.

10.5 Canning

Canning includes sealing meat in a container and then heating it to destroy all microbes capable of meat and meat products spoilage. Under normal circumstances, canned products can safely be stored at room temperature for an unspecified period.

10.6 Drying

Oldest known technique of meat preservation. Drying removes moisture from meat products, lowers the water activity (*aw*) significantly so that microbes cannot grow. Freeze-dried meats, dry sausages, and jerky products are all examples of dried meats capable of being stored at room temperature without fast spoilage.

10.7 Irradiation

Irradiation, or radiation, is a pasteurization technique achieved by exposing the meat to amounts of radiation. Irradiated fresh meat products still need cooling and wrapper to prevent spoilage, but the chilled storage life of these products is highly lengthy.

10.8 Fermentation

One early form of food preservation used in meat production is fermentation. Fermentation comprises the addition of confirmed safe bacteria to meat. These fermenting bacteria produce acid as they grow, lowering the pH of the meat and preventing the growth of several pathogenic microbes.

10.9 Vacuum packaging

Oxygen is essential for various bacteria to grow. For this purpose, most meats are vacuum-packaged, which extends the storage life undercooled circumstances to about 100 days. Besides, vacuum packaging reduces the oxidation of unsaturated fatty acids and slows the development of rancid meat.

11. Conclusion

In summary, the main sources of pathogens in meat and meat products are; the animal itself; human handlers; equipment's in contact, environmental sources or water used in preparation. Meat and its products have been engaged in many diseases or outbreaks in human consumers which necessity awareness and educational knowledge about causative agents and control hygienic measures. Therefore, strict hygienic precautions must me prevailed during meat handling and preparations.

Author details

Dhary Alewy Almashhadany
Department of Medical Lab Science (DMLS), College of Science (CSCN), Knowledge University (KNU), Erbil, Kurdistan Region, Iraq

*Address all correspondence to: dhary.hammed@knu.edu.iq

References

[1] Boler, D.D. and Woerner, D.R. (2017): What is meat? A perspective from the American Meat Science Association, Animal Frontiers, Volume 7, Issue 4, October 2017, Pages 8-11, https://doi.org/10.2527/af.2017.0436.

[2] O'Connor LE, Gifford CL, Woerner DR, Sharp JL, Belk KE, Campbell WW. Dietary Meat Categories and Descriptions in Chronic Disease Research Are Substantively Different within and between Experimental and Observational Studies: A Systematic Review and Landscape Analysis [published correction appears in Adv Nutr. 2020 Jan 1;11(1):180]. Adv Nutr. 2020;11(1):41-51. doi:10.1093/advances/nmz072.

[3] Listrat, A., Lebret, B., Louveau, I.,Astruc, T. , Bonnet, M., Lefaucheur, L., Picard, B.and Bugeon, J. (2016): How Muscle Structure and Composition Influence Meat and Flesh Quality. The Scientific World Journal, vol. 2016, Article ID 3182746, 14 pages, 2016. https://doi.org/10.1155/2016/3182746.

[4] Obeid R, Heil SG, Verhoeven MMA, van den Heuvel EGHM, de Groot LCPGM, Eussen SJPM. Vitamin B12 Intake from Animal Foods, Biomarkers, and Health Aspects. Front Nutr. 2019; 6:93. Published 2019 Jun 28. doi:10.3389/fnut.2019.00093.

[5] Gómez, I., Janardhanan, R., Ibañez, F.C. and Beriain, M.J. (2020): The Effects of Processing and Preservation Technologies on Meat Quality: Sensory and Nutritional Aspects.

[6] Tessari P, Lante A, Mosca G. (2016): Essential amino acids: master regulators of nutrition and environmental footprint? Sci Rep. 2016; 6:26074. Published 2016 May 25. doi:10.1038/srep26074.

[7] Fayet, F., Flood, V., Petocz, P., & Samman, S. (2013). Avoidance of meat and poultry decreases intakes of omega-3 fatty acids, vitamin B12, selenium and zinc in young women. Journal of Human Nutrition and Dietetics, 27, 135-142. doi:10.1111/jhn.12092.

[8] Theobald, H.E. (2005): Dietary calcium and health. Nutrition Bulletin, first published: 17 August 2005 https://doi.org/10.1111/j.1467-3010.2005.00514.x.

[9] Rouger A, Tresse O, Zagorec M. (2017): Bacterial Contaminants of Poultry Meat: Sources, Species, and Dynamics. Microorganisms. 2017;5(3):50. Published 2017 Aug 25. doi:10.3390/microorganisms5030050.

[10] Zhao, X., Ren, W., Siegel, P.B., Li, J., Wang, Y., Yin, H., Zhang, Y., Lai, S., Shu, G. and Zhu, Q. (2018) Meat quality characteristics of chickens as influenced by housing system, sex, and genetic line interactions, Italian Journal of Animal Science, 17:2, 462-468, DOI: 10.1080/1828051X.2017.1363639.

[11] Abebe, E., Gugsa, G., Ahmed, M. (2020): Review on Major Food-Borne Zoonotic Bacterial Pathogens", Journal of Tropical Medicine, vol. 2020, Article ID 4674235, 19 pages, 2020. https://doi.org/10.1155/2020/4674235.

[12] Bintsis T. (2017): Foodborne pathogens. AIMS Microbiol. 2017;3(3):529-563. Published 2017 Jun 29. doi:10.3934/microbiol.2017.3.529.

[13] CDC (Centers for Disease Control and Prevention) (2020): Diseases & Conditions A-Z Index . https://www.cdc.gov/diseasesconditions/az/a.html.

[14] Belay, E.D. and Schonberger, L.B. (2005): THE PUBLIC HEALTH IMPACT OF PRION DISEASES. Annu. Rev. Public Health 2005. 26:191-212. doi: 10.1146/annurev.publhealth.26.021304.144536.

[15] Glatzel M, Stoeck K, Seeger H, Lührs T, Aguzzi A. Human Prion Diseases: Molecular and Clinical Aspects. Arch Neurol. 2005;62(4):545-552. doi:10.1001/archneur.62.4.545.

[16] Aldy, J.E. and Viscusi, W.K. (2012): Risk Regulation Lessons from Mad Cows. Foundations and Trends in Microeconomics, Vol. 8, No. 4 (2012) 231-313. DOI: 10.1561/0700000046.

[17] Espinosa, J.C., Morales, M., Herva, M.E. and Torres, j.m.(2006): Transmission of bovine spongiform encephalopathy. FUTURE VIROLOGYVOL. 1, NO. 3. https://doi.org/10.2217/17460794.1.3.393.

[18] Imran, M., Mahmood, S. An overview of human prion diseases. Virol J 8, 559 (2011). https://doi.org/10.1186/1743-422X-8-559.

[19] CDC (Centers for Disease Control and Prevention) (2018): Prevention Measures against BSE Spread.

[20] O'Shea H, Blacklaws BA, Collins PJ, McKillen J, Fitzgerald R. Viruses Associated with Foodborne Infections. Reference Module in Life Sciences. 2019; B978-0-12-809633-8.90273-5. doi:10.1016/B978-0-12-809633-8.90273-5.

[21] S'anchez, G.(2015): Processing Strategies to Inactivate Hepatitis AVirus in Food Products: A Critical Review. ComprehensiveReviewsinFood Science and FoodSafety , Vol.14. Doi: 10.1111/1541-4337.12154.

[22] Kwon SY. and Lee CH. (2011): Epidemiology and prevention of hepatitis B virus infection. Korean J Hepatol. 2011;17(2):87-95. doi:10.3350/kjhep.2011.17.2.87.

[23] Manns, M.P., Buti, M., Gane, E.D., Pawlotsky, J.M., Razavi, H., Terrault, N. and Younossi, Z. (2017): Hepatitis C virus infection. NATURE REVIEWS | DISEASE PRIMERS VOLUME 3 | ARTICLE NUMBER 17006. doi:10.1038/nrdp.2017.6.

[24] Rizzetto M. (2015): Hepatitis D Virus: Introduction and Epidemiology. Cold Spring Harb Perspect Med. 2015;5(7): a021576. Published 2015 Jul 1. doi:10.1101/cshperspect. a021576.

[25] Kamar N, Dalton HR, Abravanel F, Izopet J. Hepatitis E virus infection. Clin Microbiol Rev. 2014;27(1):116-138. doi:10.1128/CMR.00057-13.

[26] Petrović T. and D'Agostino M. (2016): Viral Contamination of Food. Antimicrobial Food Packaging. 2016;65-79. doi:10.1016/B978-0-12-800723-5.00005-X.

[27] Crawford SE, Ramani S, Tate JE, et al. Rotavirus infection. Nat Rev Dis Primers. 2017; 3:17083. Published 2017 Nov 9. doi:10.1038/nrdp.2017.83.

[28] Shrestha, S., Thakali, O., Raya, S. et al. Acute gastroenteritis associated with Rotavirus A among children less than 5 years of age in Nepal. BMC Infect Dis 19, 456 (2019). https://doi.org/10.1186/s12879-019-4092-2.

[29] Sun, X., Wang, L., Qi, J., Li, D., Wang, M., Cong, X., Peng, R., Chai, W., Zhang, Q., Wang,H., Wen, H., Gao, G.F., Tan ,M. and Duan, Z. (2018): Human Group C Rotavirus VP8*s Recognize Type A Histo-Blood Group Antigens as Ligands. Journal of Virology May 2018, 92 (11) e00442–e00418; DOI: 10.1128/JVI.00442-18.

[30] Tesson, V., Federighi, M., Cummins, E., Mota, J.O., Guillou, S. and Boué, G. (2020): A Systematic Review of Beef Meat Quantitative Microbial Risk Assessment Models. Int. J. Environ. Res. Public Health 2020, 17, 688; doi:10.3390/ijerph17030688.

[31] Almashhadany, D.A.(2021): Isolation, biotyping and antimicrobial susceptibility of Campylobacter isolates

from raw milk in Erbil city, Kurdistan Region, Iraq. Italian Journal of Food Safety (under publishing).

[32] Alaboudi, A.R., Alsawaf, S.D. and Almashhadany, D.A. (1991): Food Hygiene. Mosul University. College of Veterinary Medicine, Mosul, Iraq.

[33] Moxley, R.A., Bargar, T.W., Kachman, S.D., Baker, D.R. and Francis, D.H. (2020): Intimate Attachment of *Escherichia coli* O157:H7 to Urinary Bladder Epithelium in the Gnotobiotic Piglet Model . Microorganisms 2020, 8, 263; doi:10.3390/microorganisms 8020263.

[34] Hendriks, A.C.A., Reubsaet, F.A.G., Kooistra-Smid, A.M.D.M. et al. Genome-wide association studies of Shigella spp. and Enteroinvasive Escherichia coli isolates demonstrate an absence of genetic markers for prediction of disease severity. BMC Genomics 21, 138 (2020). https://doi.org/10.1186/s12864-020-6555-7.

[35] Lim, M.A., Kim, J.Y., Acharya, D., Bajgain, B.B., Park, J.H., Yoo, S.J. and Lee, K.(2020): A Diarrhoeagenic Enteropathogenic *Escherichia coli* (EPEC) Infection Outbreak That Occurred among Elementary School Children in Gyeongsangbuk-Do Province of South Korea Was Associated with Consumption of Water-Contaminated Food Items. International journal of environmental research and public health, 17(9), 3149. https://doi.org/10.3390/ijerph17093149

[36] WHO (World Health Organizaiton) (2019): Food Safety. Available online: https://www.who.int/news-room/fact-sheets/ detail/food-safety (accessed on 4 June 2019).

[37] Fleckenstein, J.M. and Kuhlmann, F.M. (2019): Enterotoxigenic Escherichia coli Infections. Curr Infect Dis Rep 21, 9 (2019). https://doi.org/10.1007/s11908-019-0665-x

[38] Okhuysen, P. C. and DuPont, H. L. (2010). EnteroaggregativeEscherichia coli(EAEC): A Cause of Acute and Persistent Diarrhea of Worldwide Importance. The Journal of Infectious Diseases, 202(4), 503-505. doi:10.1086/654895.

[39] Almashhadany, D.A., Abdul-Rahman Shater, A.R., Ba-Salamah, H.A. and Abd Algalil, F.M. (2018): Prevalence of *Listeria monocytogenes* in Human in Dhamar Governorate/ Yemen. Journal of Medical and Pharmaceutical Sciences Issue (1), Volume (2). DOI: 10.26389/AJSRP.F181217.

[40] Ranasinghe, R.A.S.S., Satharasinghe, D.A., Tang, J.Y.H., Rukayadi, Y., Radu, K.R., New, C.Y., Son, R. and Premarathne, J.M.K.J.K. (2021): Persistence of *Listeria monocytogenes* in food commodities: foodborne pathogenesis, virulence factors, and implications for public health. Food Research 5 (1) (2021) 1-16. DOI: https://doi.org/10.26656/fr.2017.5(1).199.

[41] Almashhadany, D. A. and Osman, A. A. (2019). Isolation, Serotyping, and Antibiogram of Salmonella Isolates from Raw Milk Sold at Retail Vending in Erbil City, Iraq. Bulletin of University of Agricultural Sciences and Veterinary Medicine Cluj-Napoca. Animal Science and Biotechnologies, 76(2), 116-122.

[42] Heredia, N. and García, S. (2018): Animals as sources of food-borne pathogens: A review. Anim Nutr. 2018;4(3):250-255. doi:10.1016/j.aninu.2018.04.006.

[43] PHAC (Public Health Agency of Canada) (2020): A-Z infectious diseases - Canada.ca. Infectious diseases. https://www.canada.ca/en/public-health/services/infectious-diseases.html.

[44] Ranjbar, R.and Farahani, A. (2019): Shigella: Antibiotic-Resistance Mechanisms And New Horizons For

Treatment. Infect Drug Resist. 2019; 12:3137-3167. https://doi.org/10.2147/IDR.S219755.

[45] Al-Dahmoshi, H.O.M., Al-Khafaji, N.S.K., Al-Allak, M.H., Salman, W.K. and Alabbasi, A.H. (2020): A review on shigellosis: Pathogenesis and antibiotic resistance. Drug Invention Today | Vol 14 • Issue 5 • 2020.

[46] Nelson EJ, Harris JB, Morris JG Jr, Calderwood SB, Camilli A. (2009): Cholera transmission: the host, pathogen and bacteriophage dynamic. Nat Rev Microbiol. 2009;7(10):693-702. doi:10.1038/nrmicro2204.

[47] Dinede G, Abagero A, Tolosa T. (2020): Cholera outbreak in Addis Ababa, Ethiopia: A case-control study. PLoS One. 2020;15(7): e0235440. Published 2020 Jul 2. doi: 10.1371/journal.pone.0235440

[48] Gupta, V., Gulati, P., Bhagat, N., Dhar, M. S., & Virdi, J. S. (2014). Detection of Yersinia enterocolitica in food: an overview. European Journal of Clinical Microbiology & Infectious Diseases, 34(4), 641-650. doi:10.1007/s10096-014-2276-7.

[49] Chlebicz, A. and Slizewska, K. (2018): Campylobacteriosis, Salmonellosis, Yersiniosis, and Listeriosis as Zoonotic Foodborne Diseases: A Review. Int. J. Environ. Res. Public Health 2018, 15, 863; doi:10.3390/ijerph15050863.

[50] Almashhadany, D.A. (2019): The significance of milk ring test for identifying Brucella antibodies in cows and buffaloes' raw milk at Erbil governorate, Kurdistan region, Iraq. Iraqi Journal of Veterinary Sciences, Vol. 33, No. 2, 2019 (395-400). DOI: 10.33899/ijvs.2019.163085.

[51] Almashhadany, D.A. (2009): Serological Study on Brucellosis in Sheep and Goats in Thamar Province , Yemen .Egypt. J of Appl. Sci., 24 (3): 28-38.

[52] Almashhadany, D.A. (2018): Zoonotic Diseases / Part Two / Bacterial Zoonoses 1- Brucellosis. LAP LAMBERT Academic Publishing (October 12, 2018) ISBN-10: 3659823023.

[53] Almashhadany, D.A. (2014): Prevalence of Brucellosis in Human and Camels in Thamar Province / Yemen. J. Saudi Soc. For Agric. Sci , Vol 13(1a): 108-132.

[54] Almashhadany, D.A. (2018): The Utility of MRT to Screen Brucellosis among Ewe and Nanny Goats Milk in Erbil Governorate / Kurdistan Region / Iraq. IJBPAS, September, 2018, 7(9): 1786-1802. https://doi.org/10.31032/IJBPAS/2018/7.9.4551.

[55] Hashimi, M.H., Hashimi, R. and Ryan, Q. (2020): Toxic Effects of Pesticides on Humans, Plants, Animals, Pollinators and Beneficial Organisms. APRJ, 5(4): 37-47, 2020; Article no. APRJ.59338. DOI: 10.9734/APRJ/2020/v5i430114.

[56] Silva, C.C.P., Zannin, M., Rodrigues, D.S., Santos, C.R., Correa, I.A. and Junior, V.H. (2010): Clinical and Epidemiological Study of 27 Poisonings Caused by Ingesting Puffer Fish (Tetrodontidae) in the States of Santa Catarina and Bahia, Brazil. Rev. Inst. Med. trop. S. Paulo, 52(1):51-55, January–February, 2010. doi: 10.1590/S0036-46652010000100010.

[57] Wells, K., Butterworth, A. and Richards, N. (2020): A review of secondary pentobarbital poisoning in scavenging wildlife,companion animals and captive carnivores. J Vet Forensic Sci, jvfs.net, January 2020, Vol. 1, No. 1.

[58] Petersen LE., Kellermann M.Y., Schupp P.J. (2020): Secondary Metabolites of Marine Microbes: From Natural Products Chemistry to

Chemical Ecology. In: Jungblut S., Liebich V., Bode-Dalby M. (eds) YOUMARES 9 - The Oceans: Our Research, Our Future. Springer, Cham. https://doi.org/10.1007/978-3-030-20389-4_8.

[59] Abdulmumeen, H.A. Risikat, A.N. and Sururah, A.R.(2012): Food: Its preservatives, additives and applications. International Journal of Chemical and Biochemical Sciences, 1(2012): 36-47. DOI: 10.13140/2.1.1623.5208.

[60] Almashhadany D.A. (2020): Monitoring of antibiotic residues among sheep meats at Erbil city and thermal processing effect on their remnants. Iraqi Journal of Veterinary Sciences, Vol. 34, No. 2, 2020 (217-222) 217. DOI: 10.33899/ijvs.2019.125814.1161.

[61] CDC (Centers for Disease Control and Prevention) (2017): National Center for Emerging and Zoonotic Infectious Diseases (2017): How Food Gets Contaminated - The Food Production Chain.

[62] Watari, T., Tachibana, T., Okada, A., Nishikawa, K., Otsuki, K., Nagai, N., Abe, H., Nakano, Y., Takagi, S. and Amano. Y. (2020): A review of food poisoning caused by local food in Japan. J Gen Fam Med. 2020; 00:1-9. DOI: 10.1002/jgf2.384.

[63] Jessberger, N., Dietrich, R., Granum, P.E. and Märtlbauer, E. (2020): The Bacillus cereus Food Infection as Multifactorial Process. Toxins 2020, 12, 701; doi:10.3390/toxins12110701.

[64] Gharib AA, El-Hamid MIA, El-Aziz NKA, Yonan EY, Allam MO (2020). *Bacillus cereus*: Pathogenicity, viability and adaptation. Adv. Anim. Vet. Sci. 8(s1): 34-40. DOI | http://dx.doi.org/10.17582/journal.aavs/2020/8.s1.34.40.

[65] Tewari A, Abdullah S. Bacillus cereus food poisoning: international and Indian perspective. J Food Sci Technol. 2015 May;52(5):2500-11. doi: 10.1007/s13197-014-1344-4. Epub 2014 Apr 13. PMID: 25892750; PMCID: PMC4397285.

[66] Zeighami, H., Nejad-dost, G., Parsadanians, A., Daneshamouz, S. and Haghi, F. (2020): Frequency of hemolysin BL and non-hemolytic enterotoxin complex genes of *Bacillus cereus* in raw and cooked meat samples in Zanjan, Iran. Toxicology Reports 7 (2020) 89-92. https://doi.org/10.1016/j.toxrep.2019.12.006.

[67] Gaware, V.M., Kotade, K.B., Dolas, R.T., Dhamak,K.B. , Somawanshi, S.B. and Nikam, V.K. (2011): Botulism Foodborne Disease: A Review. J. Chem. Pharm. Res., 2011, 3(1):84-92.

[68] Kadariya J, Smith TC, Thapaliya D. (2014): Staphylococcus aureus and staphylococcal food-borne disease: an ongoing challenge in public health. Biomed Res Int. 2014; 2014:827965. doi:10.1155/2014/827965

[69] Abril, A.G., Villa, T.G., Barros-Velázquez, J., Cañas,B., Sánchez-Pérez , A., Calo-Mata, P. and Carrera, M. (2020): Staphylococcus aureus Exotoxins and Their Detection in the Dairy Industry and Mastitis.Toxins 2020, 12, 537; doi:10.3390/toxins12090537

[70] Almashhadany, D.A. (2020): ELISA-based monitoring of Toxoplasma gondii among retail sheep meat in Erbil Governorate, Kurdistan region, Iraq. Malaysian Journal of Microbiology, Vol 16(3) 2020, pp. 229-234. DOI: http://dx.doi.org/10.21161/mjm.190571

[71] Almashhadany, D.A. (2020): Survey of Toxoplasma gondii antibodies in retail red meat samples in Erbil governorate, Kurdistan Region, Iraq. SVU-IJVS, 3 (2): 51-59. DOI: 10.21608/svu.2020.31892.1057.

[72] Fayer R, Esposito, D.H. and Dubey, J.P. (2015): Human Infections with Sarcocystis Species. Clinical Microbiology Reviews Feb 2015, 28 (2) 295-311; DOI: 10.1128/CMR.00113-14.

[73] Ahmed, S.A. and Karanis, P.(2020): Cryptosporidium and Cryptosporidiosis: The Perspective from the Gulf Countries. Int. J. Environ. Res. Public Health 2020, 17, 6824; doi:10.3390/ijerph17186824.

[74] Baxby, D. and Hart, C.A. (1986): The incidence of cryptosporidiosis: a two-year prospective survey in a children's hospital. J. Hyg., Carnb. (1986), 96, 107-111.

[75] Feng Y. and Xiao L. (2017): Molecular Epidemiology of Cryptosporidiosis in China. Front Microbiol. 2017; 8:1701. Published 2017 Sep 6. doi:10.3389/fmicb.2017.01701.

[76] Hunter, P.R. and Nichols, G. (2002): Epidemiology and Clinical Features of Cryptosporidium Infection in Immunocompromised Patients. CLINICAL MICROBIOLOGY REVIEWS,0893-8512/02/$04.000 DOI: 10.1128/CMR.15.1.145-154.2002. Jan. 2002, p. 145-154.

[77] Gebrie, M. and Engdaw, T.A. (2015): Review on Taeniasis and Its Zoonotic Importance. European Journal of Applied Sciences 7 (4): 182-191, 2015. DOI: 10.5829/idosi.ejas.2015.7.4.96169.

[78] Bruschi, F. and Murrell, K.D. (2002): New aspects of human trichinellosis: the impact of new Trichinella species. Postgrad Med J. 2002;78(915):15-22. doi:10.1136/pmj.78.915.15.

[79] Dick, T. A. (2007). Diphyllobothriasis: The Diphyllobothrium latumHuman Infection Conundrum and Reconciliation with a Worldwide Zoonosis. Food-Borne Parasitic Zoonoses, 151-184. doi:10.1007/978-0-387-71358-8_4.

[80] Aibinu, I.E., Smooker, P.M. and Lopata, A.L. (2019): Anisakis Nematodes in Fish and Shellfish- from infection to allergies. International Journal for Parasitology: Parasites and Wildlife, Volume 9, August 2019, Pages 384-393. https://doi.org/10.1016/j.ijppaw.2019.04.007.

[81] Fuehrer, HP., Igel, P. and Auer, H. (2011): Capillaria hepatica in man—an overview of hepatic capillariosis and spurious infections. Parasitol Res 109, 969-979 (2011). https://doi.org/10.1007/s00436-011-2494-1.

[82] Murell, K.D. and Pozio, E. (2017): The Liver Flukes: *Clonorchis sinensis*, Opisthorchis spp, and Metorchis spp. In: J.B. Rose and B. Jiménez-Cisneros, (eds) Global Water Pathogen Project. http://www.waterpathogens.org (Robertson, L (eds) Part 4 Helminths). http://www.waterpathogens.org/book/liver-flukes Michigan State University, E. Lansing, MI, UNESCO.

[83] Chai, J.Y. and Jung, B.K. (2017):Fishborne zoonotic heterophyid infections: An update. Food and Waterborne Parasitology,Volumes 8-9, 2017, Pages 33-63. https://doi.org/10.1016/j.fawpar.2017.09.001.

[84] Li, F., Liu, X.H., Ge, H.L., Xie, C.Y., Cai, R.Y., Hu, Z.C., Zhang, Y.G. and Wang, Z.J. (2018): The discovery of Clinostomum complanatum metacercariae in farmed Chinese sucker, Myxocyprinus asiaticus. Aquaculture, Volume 495, 1 2018, Pages 273-280. https://doi.org/10.1016/j.aquaculture.2018.05.052.

[85] Menconi, V., Manfrin, C., Pastorino, P., Mugetti, D., Cortinovis, L., Pizzul, E., Pallavicini, A. and Prearo, M. (2020): First Report of Clinostomum complanatum (Trematoda: Digenea) in European Perch (Perca fluviatilis) from

an Italian Subalpine Lake: A Risk for Public Health?. Int. J. Environ. Res. Public Health 2020, 17, 1389; doi:10.3390/ijerph17041389.

[86] Alshannaq A. and Yu JH. (2017): Occurrence, Toxicity, and Analysis of Major Mycotoxins in Food. Int J Environ Res Public Health. 2017;14(6):632. Published 2017 Jun 13. doi:10.3390/ijerph14060632.

[87] Paterson RRM. (2019): Editorial for the Special Issue: Human Pathogenic Filamentous Fungi from Food/Water and Mycotoxins from Water. Microorganisms 2019, 7(1), 21; https://doi.org/10.3390/microorganisms7010021.

[88] Munkvold G.P. (2017): Fusarium Species and Their Associated Mycotoxins. In: Moretti A., Susca A. (eds) Mycotoxigenic Fungi. Methods in Molecular Biology, vol 1542. Humana Press, New York, NY. https://doi.org/10.1007/978-1-4939-6707-0_4.

[89] Benedict K, Chiller TM. and Mody RK. (2016): Invasive Fungal Infections Acquired from Contaminated Food or Nutritional Supplements: A Review of the Literature. Foodborne Pathog Dis. 2016;13(7):343-349. doi:10.1089/fpd.2015.2108.

[90] Kumar P, Mahato DK, Kamle M, Mohanta TK, Kang SG. (2017): Aflatoxins: A Global Concern for Food Safety, Human Health and Their Management. Front Microbiol. 2017; 7:2170. Published 2017 Jan 17. doi:10.3389/fmicb.2016.02170.

[91] Bui-Klimke TR. and Wu F. (2015): Ochratoxin A and human health risk: a review of the evidence. Crit Rev Food Sci Nutr. 2015;55(13):1860-1869. doi:10.1080/10408398.2012.724480.

[92] Erdoğan, A., Ghimire, D., Gürses, M., Çetin, B., Baran, A. (2018): Patulin Contamination in Fruit Juices and Its Control Measures. European Journal of Science and Technology No. 14, pp. 39-48, December 2018. DOI: 10.31590/ejosat.434750.

[93] Hueza IM, Raspantini PC, Raspantini LE, Latorre AO, Górniak SL. (2014): Zearalenone, an estrogenic mycotoxin, is an immunotoxic compound. Toxins (Basel). 2014;6(3):1080-1095. Published 2014 Mar 13. doi:10.3390/toxins6031080.

[94] Porter DL, Levine BL, Kalos M, Bagg A, June CH. Chimeric antigen receptor-modified T cells in chronic lymphoid leukemia. N Engl J Med. 2011 Aug 25;365(8):725-33. doi: 10.1056/NEJMoa1103849. Epub 2011 Aug 10. Erratum in: N Engl J Med. 2016 Mar 10;374(10):998. PMID: 21830940; PMCID: PMC3387277.

[95] Cenci-Goga, B.T., Iulietto, M.F., Sechi, P., Borgogni, E., Karama, M. and Grispoldi, L. (2020): New Trends in Meat Packaging. Microbiol. Res. 2020, 11, 56-67; doi:10.3390/microbiolres11020010.

Chapter 11

Effects of Pre-Slaughter Stress on Meat Characteristics and Consumer Experience

Bruno I. Cappellozza and Rodrigo S. Marques

Abstract

The current concern regarding how animals are raised, which kind of feedstuffs were fed, and the management activities employed in the livestock segment system is increasing, primarily due to the public and/or customer opinion. Therefore, a positive pressure is being placed in the industry/production to be more effective in communicating these processes and to explain what indeed occurs during the animal's productive life, from birth to slaughter. Hence, it is imperative to explain what type of situations animals face during their productive lives and how these might impact productive, health, and the quality of the final product sold at the supermarket. Additionally, it is important to understand that technologies have been developed that could mitigate some of these stress-related losses (health and productive), as well as to improve meat quality traits and overall customer eating experience.

Keywords: cattle, customer experience, DFD, nutrition, pre-slaughter stress

1. Introduction

During daily management activities, beef and dairy animals are exposed to several situations that may trigger a stress-induced inflammatory response. This response, in turn, might greatly impact health, performance, and well-being of the herd, which affects the overall profitability of livestock operations. Therefore, it is paramount to understand the mechanisms underlying the occurrence of stress and how we can use technologies to alleviate the negative effects of this response for the herd. Hence, the objective of this review is to provide an overview on stress physiology, immune system, and the interaction among these, as well as its effects on meat characteristics of beef animals and consumer acceptability implications on edible products.

2. Stress definition and physiology

The term stress is classically defined as the reaction of an animal to factors that potentially influence its homeostasis, whereas animals that are unable to cope with these factors are classified as stressed [1]. As aforementioned, ruminants are inevitably exposed to several management situations that expose them to the occurrence of stress, such as vaccination, weaning, transport among farms, transport to the stockyard, feedlot, and slaughter, novel environments, management, restriction of water and feed, among others [2]. These situations, also known as stressors, can be categorized as:

Psychological stressors: Include weaning, arrival at a novel environment, commingling.

Physical stressors: Include castration, bruises resulting from fights within a feedlot pen and/or other animals mounting, vaccination abscesses, dehorning.

Physiological stressors: Include endocrine and metabolic alterations resulting from psychological and physical stressors.

Typically, animals are exposed to various stressors throughout the production cycle that elicit a stress response, which may not involve a viral/bacterial/fungi/protozoa pathogen. Nonetheless, it is important to note that the stress may increase the susceptibility of the animal to an individual or group of pathogen(s). Therefore, it is worth mentioning that these pathogens, in turn, might be already living in the organism of the animal, but after the beginning of a stress-induced inflammatory response, the pathogen(s) is(are) able to act and trigger its effects. Although indispensable and needed for the resumption of homeostasis, stress-induced response, and its upcoming inflammatory cascade may be unnecessary and have negative effects on performance and health of the herd.

From a physiological standpoint, when an animal perceives a stressor, the immediate response is the activation of the hypothalamic–pituitary–adrenal (HPA) axis [3], characterized by the synthesis and release of corticotropin-releasing hormone (CRH) and vasopressin (VP) by their respective neurons located in the hypothalamus into the paraventricular nucleus [2]. In cattle, CRH has more potent stimulatory actions than VP [4] in a manner that upon its binding to membrane receptors in the pituitary gland, protein kinase-A is activated and 3′,5′ cyclic AMP (cAMP) is produced, leading to a calcium influx that will activate the release of adrenocorticotropic hormone (ACTH) by the pituitary gland [5]. Within the anterior pituitary gland, corticotrophs are responsible for the production of ACTH and its main function is to promote cholesterol uptake, as well as synthesis and release of steroids by the adrenal gland [2].

The adrenal gland is divided into the cortex and medulla, whereas the cortex is responsible for the synthesis and release of 3 hormone types: mineralocorticoids, androgens, and glucocorticoids. In humans and most mammalian species, cortisol is the primary glucocorticoid [2] and generally classified as the "stress hormone". In the metabolism, cortisol elicits several important functions, such as (i) breakdown of glycogen, muscle, and adipose tissues as a mechanism to provide energy to the host during an immunological challenge, (ii) hepatic production of acute-phase proteins (APP), (iii) regulation of stress response, (iv) greater synthesis and release of catecholamines, and (v) suppression of the inflammatory and immune systems to prevent autoimmune disorders. Hence, in situations where chronic diseases or stress are observed, cortisol remains elevated for a prolonged period of time and promotes an anti-inflammatory and immunosuppressive response by decreasing the synthesis and release of pro-inflammatory cytokines [6, 7]. Conversely, acute increases in cortisol concentrations have been reported to trigger a transient and temporary inflammatory response. In ruminants, greater cortisol concentrations have been associated with reduced growth rates and reproductive performance in beef females [8–10].

3. How does the immune system react to the stress?

Before entering into the specifics of the link between immunity and stress, it is worth mentioning how the immune system is structured and its main features.

The immune system is divided into innate and acquired immunity. The acquired immunity is responsible for adapting and building an immune response for each antigen the body encounters, characterized by the production of antibodies and the development of immunological memory [11]. Therefore, it does not come as a surprise that specificity is one of the main features of this branch of the immune system. On the other hand, the innate immunity is characterized by a lack of specificity, given that a similar response is observed when caused by a bacteria, protozoa, fungi, virus, or stress [2]. The barriers created by the body include (i) physical: skin, tears, and mucosal secretions, (ii) chemical: antimicrobial peptides, superoxide anion, and nitric oxide, and (iii) complement system [11]. The primary goal of the innate immune system is to provide enough time for the acquired immunity to develop a strong and effective response against any specific pathogen. The group of cells that comprise the innate immune system includes phagocytic cells (neutrophils, monocytes, macrophages, and dendritic cells), natural killer cells, and cells that release inflammatory mediators, such as mast cells, basophils, and eosinophils.

The innate immunity recognizes certain structures present in different microorganisms, known as pathogen-associated molecular patterns (PAMPs). This is the reason that lipopolysaccharide, a component of the cell wall of gram-negative bacteria, causes an immune reaction in several species, leading to an increase in cortisol concentrations and an acute-phase reaction in cattle [12]. When PAMPs interact with toll-like receptors (TLR) inside or on the surface of phagocytic cells, a cytokine response is initiated in neutrophils and macrophages by the activation of a transcription factor called nuclear factor kappa beta (NFκB), which induces the expression of genes part of the innate immunity, such as cytokines, chemokines, and co-stimulatory molecules [11].

Cytokines are chemical messengers released by phagocytic cells during an immune response and act as mediators of intermediary metabolism in immune-challenged animals [13]. The major pro-inflammatory cytokines released by cells and recognized as important players in an immune response include interleukin-6 (IL-6), IL-1 and tumor necrosis factor-α (TNF-α) [14], which their plasmatic concentrations increase following an acute bacterial challenge (LPS) [12]. Following the injury caused by a pathogenic challenge, the body builds a specific and complex set of reactions aiming to destroy the infectious pathogen, to prevent further tissue damage, and to restore homeostasis. These early and immediate set of responses are called 'inflammatory responses' and are part of the acute-phase response (APR) [15]. Tissue macrophages and blood monocytes are the major cell components involved in the APR and after activation, these cells release IL-1 and TNF-α in the circulation [16]. During an APR, several metabolic changes are observed, but the main physiological responses are increased body temperature (febrile response) and alterations in liver metabolism. These responses are primarily regulated by the aforementioned pro-inflammatory cytokines (IL-1, IL-6, and TNF-α).

Febrile response: a defense mechanism developed to control replication, growth, and kill different pathogens by preventing the formation of bacterial coats. It is mainly induced when the eicosanoid prostaglandin-E2 (PGE2) is produced from 20 carbon omega-6-derived fatty acids. Hence, it is imperative to mention that this response may be modulated by the fatty acid profile in the diets of the animals. In other words, diets containing a greater concentration of omega-6 fatty acids (i.e., linoleic and its derivatives) will lead to a greater pro-inflammatory state, whereas diets containing greater amounts of omega-3 fatty acids (i.e., linolenic and its derivatives) will promote an anti-inflammatory response [17, 18]. Although imperative for controlling an infectious challenge, the increase in body temperature does not come without a significant effect on nutrient requirements of the animals. More specifically, for every 1 °C increase in body temperature associated with an immune response, it is estimated that energy requirements are increased by 10–13% [19].

Hepatic metabolism: under homeostasis, hepatocytes produce several acute-phase proteins (APP) at a relatively steady state, but this manufacturing state dramatically changes when the animal is immunologically challenged. Noteworthy is the fact that not only the amount of APP is significantly changed during an immune response, but also the APP profile is changed, given that some increase and other decrease following an immune challenge [20]. The increase in liver metabolism also leads to dramatic increases in metabolizable protein (MP) requirements in ruminants. Moriel et al. (2015) demonstrated that increasing MP supply from 85 to 115% of daily requirements to newly-weaned beef calves improved performance during a 42-day preconditioning period, without effects on messenger RNA expression of hepatic genes involved in the production of two of the major APP, haptoglobin and ceruloplasmin [21]. Moreover, some APP peak between 1–4 days post-challenge, reducing dry matter intake (DMI) and animal performance, as well as increasing the incidence of morbidity and antimicrobial treatment [22–24]. In ruminants, haptoglobin, ceruloplasmin, serum amyloid-A, and fibrinogen have been the most studied APP [25], with significant attention given to haptoglobin and its effects when evaluating an APR. Haptoglobin has been the most reliable and consistent APP because it is almost undetectable in healthy animals, whereas a significant increase is observed after a disease, injury, or stress response [23, 26, 27].

Nonetheless, PAMP recognition may not apply in stress situations as no pathogen is directly involved at the beginning of the stress-induced inflammatory response and the cortisol is responsible for triggering such response and tissue mobilization [27, 28]. Additionally, in a stressful situation, damage-associated molecular patterns (DAMPs) are host biomolecules that can initiate and perpetuate a non-infectious inflammatory response [29]. Several DAMPs are nuclear or cytosolic proteins with a defined intracellular function that, when released outside the cell after injury and/or mobilization, move from a reducing to an oxidizing milieu resulting in their functional denaturation [30]. Examples of these molecules include DNA, RNA, mono- and polysaccharides, purine metabolites (ATP, ADP, and uric acid), as well as S-100 proteins [31–33]. In a research effort to evaluate the mechanisms underlying the effects of stress on inflammation, Cooke and Bohnert (2011) developed a neuroendocrine stress model using CRH as a non-pathogenic stressor [27]. These authors reported that cortisol peaked 1 h after CRH infusion at a dose of 0.1 μg/kg body weight (BW) followed by increases in plasma concentrations of IL-6 and haptoglobin, demonstrating stress also activates the innate immune response in animals without the presence of a specific pathogen.

4. Effects of stress on health and performance

In beef cattle, weaning is considered one of the most stressful events in the entire life of the animals. In order to evaluate this response, Arthington and colleagues (2005) evaluated the effects of early vs. traditional weaning on stress markers and growth performance of beef animals [34]. These authors reported that at the time traditional weaning occurred (approximately 300 days of age), early-weaned were lighter than traditionally-weaned cohorts (48 kg), whereas ADG (days 0–28 post-weaning = 0.87 vs. 0.40 kg/day; days 29–112 post-weaning = 1.38 vs. 1.18 kg/day) and feed efficiency (days 0–28 post-weaning = 157 vs. 81 g/kg; days 29–112 post-weaning = 159 vs. 136 g/kg) were greater for early-weaned calves. From traditional weaning until slaughter, early-weaned animals gained roughly 30 kg more BW and had greater feed efficiency than traditionally-weaned calves.

These results can be explained by the relative steady concentrations of haptoglobin and ceruloplasmin in early-weaned calves at the moment traditional weaning occurred and a lessened APR observed in early-weaned steers receiving a pathogenic challenge, suggesting that the animal might be less susceptible to developing any kind of disease following a period in which its immune system might be suppressed upon facing a stressful situation.

Another set of stressful events is transport and feedlot entry, which may occur together or in a short period of time. During transport, animals remain without feed and water for a significant period of time, as no rest stops are adopted in traditional beef-producing countries. In an effort to evaluate the metabolic effects that feed and water restriction cause in the ruminant, Marques and colleagues (2012) reported a similar loss in performance, as well as similar concentrations of non-esterified fatty acids (NEFA) and APR of nutrient-restricted vs. transported animals, demonstrating that feed and water deprivation are major contributors to tissue mobilization, APR, and reduced receiving feedlot performance in animals enrolled to long-distance transport [35]. In a subsequent study, the same research group [36] demonstrated that 24-hours feed restriction was considered the major factor causing a neuroendocrine response, explained by the greater cortisol, NEFA, and ceruloplasmin concentrations compared to animals water- and feed + water-restricted [36].

Based on these results, one could speculate that providing water and feed to cattle during the transport to the feedyard would alleviate the stress-induced inflammatory response of the herd. Cooke and colleagues (2013) designed a study to evaluate whether 2-hours rest stops every 8 hours of transport with full access to feed and water would benefit health and, consequently, improve performance of beef animals during feedlot receiving [37]. In partial agreement with the hypothesis, rest stops for feed and water consumption did reduce plasma cortisol, NEFA and haptoglobin concentrations, but did not improve feed intake and performance when compared to animals continuously transported for 24 hours. These results might be explained by the fact that (i) 2 hours rest stops were not enough to improve performance and (ii) unloading and loading the animals during rest stops also cause a stress response, similarly impairing performance of the herd.

Overall, these stressful events are faced by most, if not all, ruminants during their productive lives. Depending on production system and management, these events are faced more than once in their lifetime and which might negatively impact the overall productivity in beef cattle system.

5. Pre-slaughter stress

The stress animals face prior to slaughter has been under public scrutiny for a long period of time in the U.S. and other parts of the world. Public opinion has been forcing the industry to adopt good management and transport practices, demanding information on the origin of the food and how the animals were raised throughout their productive lives, with a special focus on animal welfare [38]. Therefore, it is imperative to understand the factor(s) that may cause a stress-induced response in finishing animals and how these situations impact the quality of the edible product (i.e., meat) and customer acceptability.

Several stressful situations are faced by an animal from the feedlot until slaughter, including management in the working chute, truck loading at the feedlot [37], transport itself plus feed and water deprivation [35, 36], truck unloading at the slaughter plant, weather conditions, novel environment and management, as well as feed deprivation until slaughter [39]. These situations, which may occur together or in a short period of time, also trigger a stress-induced inflammatory response that, ultimately, can impact carcass characteristics and, consequently, visual and qualitative aspects of the meat being offered in the market [40]. More specifically, animals exposed to a greater number and/or magnitude of physical and psychological stressors prior to slaughter are more prone to produce the dark, firm, and dry (DFD) carcass and this process will be covered herein.

All pre-slaughter stressors can, inevitably, alter meat quality and customer acceptance, particularly due to an increase in meat pH and changes in meat tenderness and color [41]. Meat pH is a result of the amount of pre-slaughter glycogen levels present in the muscles, which, in turn, depends greatly on the factors responsible for physical and psychological stress [42]. Exposure to stressors before slaughter results in muscle energy reduction, depletion of glycogen and changes in important meat physical and chemical attributes, such as pH, softness, and color [43]. Meat pH values greater than 6.0 at 12 to 48 h post-mortem results in DFD cuts that are more susceptible to microbial contamination and a shorter shelf-life. Other authors have also considered 5.80 as the threshold for the determination of DFD [44, 45]. High meat pH will lead to a product not appreciated and accepted by customers, as DFD cuts are dark-red to brown-black and have a dry, firm, and sticky consistency [46]. Meat traits that have greater influence on consumer satisfaction are tenderness, juiciness, and flavor of the cooked meat, all of which are severely impacted in DFD meat cuts [47].

Evaluating pre-slaughter stress and its effects on carcass characteristics, Carrasco-Garcia and colleagues (2020) reported that 80% of the carcasses analyzed had a pH greater than 5.80 [45]. The measurement of pH is one of the most important quality traits as it is related to depletion of muscle glycogen reserves [45]. After animal death by exsanguination, the muscle tissue becomes anoxic, triggering the anaerobic glycolysis cascade. High levels of stress hormones before or during slaughter decrease muscular glycogen reserves, as glycogen is hydrolyzed to lactic acid. In fact, higher pre-slaughter lactate concentrations are associated with a reduced consumer eating quality score [48]. Therefore, it is paramount that meat pH decreases from approximately 7.0 to 5.5 in order to avoid bacterial growth in the edible product [49]. As cortisol is one of the key players controlling the stress response and meat acidification, it would be reasonable to evaluate its concentration prior to slaughter. In the same study, Carrasco-Garcia et al. (2020) demonstrated that 67% of sampled animals had blood cortisol level greater than 45 ng/mL, which is indicative of excessive stressful conditions prior to slaughter, but no association has been observed among pH and blood cortisol [45].

Alterations in meat color are one of the most pronounced effects of high meat pH and, consequently, DFD occurrence. Carcasses with pH greater than 6.0 are usually classified as darker, less red, and less yellow [50]. During DFD episodes, a muscle absorbs light and meat becomes darker, which can be attributed to a lower amount of free water at its surface and lower oxygenation of myoglobin [51]. Conversely to blood cortisol levels and meat pH, Carrasco-Garcia and colleagues (2020) reported a positive association among meat pH and colorimetric parameters of crossbred beef animals [45]. Color is the most important sensory attribute that influences customer purchasing decisions, as they associate a bright cherry red color with freshness and quality, whereas any deviation from this parameter is perceived as poorer quality. Hence, a darker meat would be less acceptable by the customers and, consequently, cause substantial economical losses to the meat industry worldwide. In the early 2000s, it was estimated that the Australian beef industry lost roughly 38.5 million dollars due to DFD, resulting in an average loss of $ 90/carcass. Nonetheless, these numbers might be even greater in other regions, depending on cattle breeds and production settings in which animals are reared for fattening.

In agreement with the carcass-reduced quality cascade, meat tenderness is often observed in carcasses with greater pH and reduced lightness [52]. The mechanisms underlying a greater shear force in DFD cuts might be related to a reduced sarcomere length [53], which is recognized as an important cause of increased toughness in meat and its length increases as pH decreases below 6.2 [54]. The sarcoplasmic reticulum is a membranous cellular organelle responsible for regulating the amount of calcium ions in the sarcoplasm of the muscle fiber [55]. After slaughter, calcium concentration in the sarcoplasm increases due to loss of the calcium-accumulating ability of the sarcoplasm reticulum [56] accompanied by a gradual leakage of calcium ions into the sarcoplasm. Calcium is paramount in meat tenderization and the calpain system requires the presence of calcium to be activated [57]. Hence, relevant changes in sarcomere length begin to occur early postmortem when sarcomeres contract as the muscle goes into rigor. If calcium is released when ATP is still available, muscle contraction occurs, resulting in shorter sarcomeres and detrimental effects on meat tenderness [55]. Hence, it is possible that the amount of stress during pre-slaughter might affect calcium release and sarcomere length, resulting in a tougher meat.

Stress itself is not the only factor predisposing DFD occurrence in ruminants, in a manner that sex, breed (*Bos taurus* vs. *Bos indicus*), nutrition, animal category (bull vs. steer), temperament, and age are other important factors. In Mexico, Loredo-Osti and colleagues (2018) reported that 13.5% of the *Bos indicus*-crossbred carcasses processed by a slaughter plant were classified as DFD [58], whereas this value approached 2% in a recent survey conducted in the U.S. [59], but no data have been reported for Brazil, where zebu breeds prevail. Although no data have been reported, *Bos indicus* breeds are well-known as more temperamental than *Bos taurus* breeds [60, 61]. Hence, these animals have a heightened stress and APR response, which leads us to speculate that DFD occurrence would be greater in *B. indicus* vs. *B. taurus* herds. Besides variation among breeds, within breeds variation on temperament might also impact the stress response and carcass traits. Francisco and colleagues (2015) demonstrated that more temperamental *Bos indicus* animals had a greater carcass bruise count, reduced color index, fat content, and marbling score [62].

In summary, DFD cuts can be recognized by the customers as:

Colorimetric aspects: dark-red to brown-black

Consistency: dry, firm, and sticky

Greater tenderness and reduced juiciness

Reduced visual acceptability, desire to purchase, and customer eating experience

Hence, it is imperative to develop strategies that reduce the magnitude of a stress response and, consequently, improve carcass quality and customer sensory eating quality score. Loudon and colleagues (2019) suggested that 2-weeks resting periods following social/group mixing as an alternative to improve customer eating quality [48]. Recently, Cappellozza and colleagues (2020) reported that the administration of bovine appeasing substance (BAS; 5 mL/head) immediately at loading to slaughter reduced mean carcass pH (5.82 vs. 5.75) and the proportion of carcass classified as having post-mortem pH greater than 5.80 (42.2. vs. 26.2%) [63]. The use of appeasing pheromones has been initially discovered in swine and shown to reduce the agonistic behavior of piglets [64, 65]. Pheromones are species-specific semiochemicals compounds released from one specific individual to induce both a behavioral and physiological response in a conspecific [66]. In cattle, BAS is based on a mixture of fatty acids, reproducing the composition of the natural substance produced by the dam during the parturition [67]. Therefore, BAS is expected to have calming effects that, in turn, will improve health and performance of the herd. As an example, Cooke et al. (2020) demonstrated that BAS administration at weaning reduce mean plasma haptoglobin concentration 15 days post-weaning and improved 45-day post-weaning performance (+ 70 grams/day) [68]. Moreover, Cappellozza and colleagues (2020) reported a similar improvement in performance of pure *Bos indicus* newly-weaned grazing beef animals during a 45-day post-weaning period [63]. The benefits of BAS administration seem to rely on transient temperamental changes [69], stress-induced inflammatory response [68, 69], reduced disease susceptibility due to an increased vaccine efficacy [69], recovery from a pathogen challenge [70], and a subsequent improvement in feed efficiency [71] and BW change and gain [72].

6. DFD cuts and human health

As aforementioned, DFD cuts are undesirable because they are esthetically unpleasant and are more susceptible to microbial growth [73], but the eating quality of these cuts were less defined in beef cuts. As an example, until 1997, the United States Standards for Grades of Carcass Beef stated that "there is little or no evidence which indicates that the dark-cutting condition has any adverse effect on palatability..." [74], contradicting what has been reported for pig cuts and similar conditions of carcass issues [75, 76]. In order to address this matter, Wulf and colleagues evaluated the effects of DFD cuts on carcass traits and beef palatability [77].

These authors demonstrated that several carcass quality attributes were significantly affected in DFD cuts, such as backfat thickness (31% less), USDA quality grade (9.5% less), intramuscular fat (29% less), pH (11% greater), and lower colorimetric readings [77]. Moreover, cooked beef palatability was substantially lower for DFD vs. normal carcasses, whereas tenderness variation and shear force were greater [77]. In the sensory panel, DFD carcasses produced a greater percentage of "tough" Longissimus steaks and a reduced percentage of "very tender" steaks vs. normal carcasses [77]. Additionally, off-flavors classifications were more frequent observed in panelists analyzing DFD vs. normal cuts.

To the best of our knowledge, no other study evaluated chemical composition of DFD vs. normal cuts. It is noteworthy mention that visual parameters are the major driver of a customer's desire to buy a piece of meat and, therefore, these cuts are well desired in the shelf of a store. Nonetheless, it can also be assumed that no harm is observed after a consumption of DFD cuts, given that human sensorial analysis have been reported [77].

7. Conclusions

Stressful situations are often faced by ruminants during their entire productive lives, from birth until slaughter. These situations predispose animals to health and performance losses which, ultimately might impact carcass quality and customer eating experience. Therefore, additional strategies and/or technologies must be developed to reduce the losses caused by stress and improve carcass traits, which will result in greater acceptability of edible products by the food chain and likely reduce the scrutiny of the public opinion regarding animal production.

Conflict of interest

The authors declare no conflict of interest.

Author details

Bruno I. Cappellozza[1]* and Rodrigo S. Marques[2]

1 Nutricorp, Araras, SP, Brazil

2 Montana State University, Bozeman, MT, USA

*Address all correspondence to: cappellozza@nutricorp.com.br

References

[1] Moberg GP. Biological response to stress: Implications for animal welfare. In: The Biology of Animal Stress: Basic Principles and Implications for Animal Welfare. 2000:1-21. CAB International, Oxon, UK. doi: 10.1079/9780851993591.001

[2] Carroll JA, Forsberg NE. Influence of stress and nutrition on cattle immunity. Vet. Clin. Food Anim. 2007:23:105-149. doi: 10.1016/j.cvfa.2007.01.003

[3] Dobson H, Smith RF. What is stress, and how does it affect reproduction? Anim. Reprod. Sci. 2000:60-61:743-752. doi: 10.1016/s0378-4320(00)00080-4

[4] Carroll JA, McArthur NH, Welsh TH. In vitro and in vivo temporal aspects of ACTH secretion: stimulatory corticotropin-releasing hormone and vasopressin in cattle. J. Vet. Med. A. Physiol. Pathol. Clin. Med. 2007:54:7-14. doi: 10.1111/j.1439-0442.2007.00908.x

[5] Link H, Dayanithi G, Gratzl M. Glucocorticoids rapidly inhibit oxytocin-stimulated adrenocorticotropin release from rat anterior pituitary cells, without modifying intracellular calcium transients. Endocrinology. 1993:132:873-878. doi: 10.1210/endo.132.2.8381078

[6] Munck A, Guyre PM, Holbrook NJ. Physiological functions of glucocorticoids in stress and their relation to pharmacological actions. Endocr. Rev. 1984:5:25-44. doi: 10.1210/edrv-5-1-25

[7] Kelley KW. Cross-talk between the immune and endocrine systems. J. Anim. Sci. 1988:66:2095-2108. doi: 10.1016/j.cotox.2017-12.003

[8] Cooke RF, Arthington JD, Austin BR, Yelich JV. Effects of acclimation to handling on performance, reproductive and physiological responses of Brahman-crossbred heifers. J. Anim. Sci. 2009:87:3403-3412. doi: 10.2527/jas.2009-1910

[9] Cooke RF, Bohnert DW, Cappellozza BI, Mueller CJ, DelCurto T. Effects of temperament and acclimation to handling on reproductive performance of Bos taurus beef females. J. Anim. Sci. 2012:90:3547-3555. doi: 10.2527/jas.2011-4768

[10] Francisco CL, Cooke RF, Marques RS, Mills RR, Bohnert DW. Effects of temperament and acclimation to handling on feedlot performance of Bos taurus feeder cattle originated from a rangeland-based cow-calf system. J. Anim. Sci. 2012:90:5067-5077. doi: 10.2527/jas.2012-5447

[11] Murphy KM, Travers P, Walport M. Janeway's Immunobiology. 2008:7: 1-108.

[12] Carroll JA, Arthington JD, Chase Jr. CC. Early weaning alters the acute-phase reaction to an endotoxin challenge in beef calves. J. Anim. Sci. 2009:87:4167-4172. doi: 10.2527/jas.2009-2016

[13] Klasing KC. Nutritional aspects of leukocyte cytokines. J. Nutr. 1988:118:1436-1446. doi: 10.1093/jn/118.12.1436

[14] Johnson RW. Inhibition of growth by pro-inflammatory cytokines: an integrated view. J. Anim. Sci. 1997:75:1244-1255. doi: 10.2527/1997.7551244x

[15] Baumann H, Gauldie J. The acute-phase response. Immunol. Today. 1994:15:74-80. doi: 10.1016/0167-5699(94)90137-6

[16] Gauldie, J. Encyclopedia of Human Biology. 1991:1:25-35.

[17] Miles EA, Calder PC. Modulation of immune function by dietary fatty acids. Proc. Nutr. Soc. 1998:57:277-292. doi: 10.1079/pns19980042

[18] Schmitz G, Ecker J. The opposing effects of n-3 and n-6 fatty acids. Prog. Lipid Res. 2008:47:147-155. doi: 10.1016/j.plipres.2007.12.004

[19] Kluger MJ, Rothenburg BA. Fever and reduced iron: their interaction as a host defense response to bacterial infection. Science 1979:203:374-376. doi: 10.1126/science.760197

[20] Murata H, Shimada N, Yoshioka M. Current research on acute phase proteins in veterinary diagnosis: an overview. Vet. J. 2004:168:28-40. doi: 10.1016/S1090-0233(03)00119-9

[21] Moriel P, Artioli LFA, Poore MH, Confer AW, Marques RS, Cooke RF. Increasing the metabolizable protein supply enhanced growth performance and led to variable results on innate and humoral immune response of preconditioning beef steers. J. Anim. Sci. 2015:93:4473-4485. doi: 10.2527/jas.2015-9238

[22] Cooke RF. Invited paper: Nutritional and management considerations for beef cattle experiencing stress-induced inflammation. Prof. Anim. Sci. 2017:33:1-11. doi: 10.15232/pas.2016-01573

[23] Carter JN, Meredith GL, Montelongo M, Gill DR, Krehbiel C, Payton ME, Confer AW. Relationship of vitamin E supplementation and antimicrobial treatment with acute-phase protein responses in cattle affected by naturally acquired respiratory tract disease. Am. J. Vet. Res. 2002:63:1111-1117. doi: 10.2460/ajvr.2002.63.1111

[24] Qiu X, Arthington JD, Riley DG, Chase Jr. CC, Phillips WA, Coleman SW, Olson TA. Genetic effects on acute phase protein response to the stresses of weaning and transportation in beef calves. J. Anim. Sci. 2007:85:2367-2374. doi: 10.2527/jas.2006-843

[25] Godson DL, Baca-Estada ME, Van Kessel AG. Regulation of bovine acute phase responses by recombinant interleukin-1B. Can. J. Vet. Res. 1995:59:249-255.

[26] Petersen HH, Nielsen JP, Heegaard PMH. Application of acute phase protein measurements in veterinary clinical chemistry. Vet. Res. 2004:35:163-187. doi: 10.1051/vetres:2004002

[27] Cooke RF, Bohnert DW. Technical note: Bovine acute-phase response after corticotropin-release hormone challenge. J. Anim. Sci. 2011:89:252-257. doi: 10.2527/jas.2010-3131

[28] Arthington JD, Eicher SD, Kunkle WE, Martin FG. Effect of transportation and commingling on the acute-phase protein response, growth, and feed intake of newly weaned beef calves. J. Anim. Sci. 2003:81:1120-1125. doi: 10.2527/2003.8151120x

[29] Seong SY, Matzinger P. Hydrophobicity: an ancient damage-associated molecular pattern that initiates innate immune responses. Nature Rev. Immunol. 2004:4:469-478. doi: 10.1038/nri1372

[30] Rubartelli A, Lotze MT. Inside, outside, upside down: Damage-associated molecular pattern molecules (DAMPs) and redox. Trends Immunol. 2007:28:429-436. doi: 10.1016/j.it.2007.08.004

[31] Gardella S, Andrei C, Ferrera D, Lotti LV, Torrisi MR, Bianchi ME, Rubartelli A. The nuclear protein HMGB1is secreted by monocytes via a non-classic, vesicle-mediated secretory pathway. EMBO Reports. 2002:3:995-1001. doi: 10.1093/embo-reports/kvf198

[32] Maverakis E, Kim K, Shimoda M, Gershwin ME, Patel F, Wilken R, Raychaudhuri S, Ruhaak LR, Lebrilla CB. Glycans in the immune system and the altered glycan theoryof autoimmunity: A critical review. J. Autoimmun. 2015:57:1-13.

[33] Roh JS, Sohn DH. Damage-associated molecular patterns in inflammatory diseases. Immune Network. 2018:18:e27. doi: 10.4110/in.2018.18.e27

[34] Arthington JD, Spears JW, Miller DC. The effect of early weaning on feedlot performance and measures of stress in beef calves. J. Anim. Sci. 2005:83:933-939. doi: 10.2527/2005.834933x

[35] Marques RS, Cooke RF, Francisco CL, Bohnert DW. Effects of twenty-four hour transport or twenty-four hour feed and water deprivation on physiologic and performance responses of feeder cattle. J. Anim. Sci. 2012:90:5040-5046. doi: 10.2527/jas.2012-5425

[36] Marques RS, Bohnert DW, de Sousa OA, Brandão AP, Schumaher TF, Schubach KM, Vilela MP, Rett B, Cooke RF. Impact of 24-h feed, water, or feed and water deprivation on feed intake, metabolic, and inflammatory responses in beef heifers. J. Anim. Sci. 2019:97:398-406. doi: 10.1093/jas/sky397

[37] Cooke RF, Guarnieri Filho TA, Cappellozza BI, Bohnert DW. Rest stops during road transport: Impacts on performance and acute-phase protein responses of feeder cattle. J. Anim. Sci. 2013:91:5448-5454. doi: 10.2527/jas.2013-6357

[38] Edwards-Callaway LN, Calvo-Lorenzo MS. Animal welfare in the U.S. slaughter industry – A focus on fed cattle. J. Anim. Sci. 2020:98:skaa040. doi: 10.1093/jas/skaa040

[39] Ferguson DM, Warner RD. Have we underestimated the impact of pre-slaughter stress on meat quality in ruminants? Meat Sci. 2008:80:12-19. doi: 10.1016/j.meatsci.2008.05.004

[40] FMI. Food Marketing Institute. Foundation for Meat & Poultry Research & Education. The power of meat, An in-depth look at meat and poultry through the hoppers' eyes. 2019. Available from: https://www.fmi.org/ [accessed 2020-11-26].

[41] Lomiwes D, Farouk MM, Wu G, Young OA. The development of meat tenderness is likely to be compartmentalized by ultimate pH. Meat Sci. 2014:96:646-651. doi: 10.1016/j.meatsci.2013.08.022

[42] Amtmann V, Gallo C, van Schaik G, Takich N. Relationships between ante-mortem handling, blood-based stress indicators, and carcass pH in steers. Arch. Med. Vet. 2006:38:529-564. doi: 10.4067/S0301-732X2006000300010

[43] Gregory NG. Animal welfare at markets and during transport and slaughter. Meat Sci. 2008:80:2-11. doi: 10.1016/j.meatsci.2008.05.019

[44] Grandin T. The effect of stress on livestock and meat quality prior to and during slaughter. Int. J. Stud. Anim. Probl. 1980:1:313-337.

[45] Carrasco-Garcia AA, Pardío-Sedas VT, León-Banda GG, Ahuja-Aguirre C, Paredes-Ramos P, Hernández-Cruz BC, Murillo VV. Effect of stress during slaughter on carcass characteristics and meat quality in tropical beef cattle. Asian-Australas. J. Anim. Sci. 2020:33:1656-1665. doi: 10.5713/ajas.19.0804

[46] Pérez-Linares C, Sánchez-López E, Ríos-Rincón FG, Olivas-Valdés JA, Figueroa-Saavedra F, Barreras-Serrano A. Pre and post slaughter cattle and carcass management

factors associated to presence of DFD beef in the hot season. Rev. Mex. Cienc. Pecu. 2013:4:149-160.

[47] Apple JK, Kegley EB, Galloway DL, Wistuba TJ, Rakes LK. Duration of restraint and isolation stress as a model to study the dark-cutting condition in cattle. J. Anim. Sci. 2005:83:1202-1214. doi: 10.2527/2005.8351202x

[48] Loudon KMW, Tarr G, Lean IJ, Polkinghorne R, McGilchrist P, Dunshea FR, Gardner GE, Pethick DW. The impact of pre-slaughter stress on beef eating quality. Animals. 2019:9:612. doi: 10.3390/ani9090612

[49] Gupta S, Earley B, Crowe MA. Effect of 12-hour road transportation on physiological, immunological and haematological parameters in bulls housed at different space allowances. Vet. J. 2007:173:605-616. doi: 10.1016/j.tvjl.2006.03.002

[50] Poleti MD, Moncaua CT, Silva-Vignatoa B, Rosa AF, Lobo AR, Cataldi TR, Negrão JA, Silva SL, Eler JP, Balieiro JCC. Label-free quantitative proteomic analysis reveals muscle contraction and metabolism proteins linked to ultimate pH in bovine skeletal muscle. Meat Sci. 2018:145:209-219. doi: 10.1016/j.meatsci.2018.06.041

[51] Tang J, Faustman C, Hoagland TA, Mancini RA, Seyfert M, Hunt MC. Postmortem oxygen consumption by mitochondria and its effects on myoglobin form and stability. J. Agric. Food Chem. 2005:53:1223-1230. doi: 10.1021/jf048646o

[52] Holdstock J, Aalhaus JL, Uttaro BA, Lopez-Campos O, Larsen IL, Bruce HL. The impact of ultimate pH on muscle characteristics and sensory attributes of longissimus thoracis within the dark cutting (Canada B4) beef carcass grade. Meat Sci. 2014:98:842-849. doi: 10.1016/j.meatsci.2014.07.029

[53] Koohmaraie M, Doumit ME, Wheeler TL. Meat toughening does not occur when rigor shortening is prevented. J. Anim. Sci. 1996:74:2935-2942. doi: 10.2527/1994.7251232x

[54] Purchas RW. An assessment of the role of pH differences in determining the relative tenderness of meat from bulls and steers. Meat Sci. 1990:27:129-140. doi: 10.1016/0309-1740(90)90061-A

[55] Ribeiro FA, Domenech-Pérez KI, Contreras-Castillo CJ, Hart K, Herrera NJ, Calkins CR. Feeding distillers grains to cattle may affect beef tenderness early postmortem. J. Anim. Sci. 2019:97:657-668. doi: 10.1093/jas/sky445

[56] Greaser M, Cassens R, Hoeskstra W, Briskey E. The effect of pH and temperature on the calcium accumulating ability of purified sarcoplasmic reticulum. J. Food Sci. 1969:34:633-637. doi: 10.1111/j.1365-2621.1969.tb12109.x

[57] Goll DE, Thompson VF, Li H, Wei W, Cong J. The calpain system. Physiol. Rev. 2003:83:731-801. doi: 10.1152/physrev.00029.2002

[58] Loredo-Osti J, Sánchez-López E, Barreras-Serrano A, Figueroa-Saavedra F, Pérez-Linares C, Ruiz-Albarrán M, Domínguez-Muñoz MA. An evaluation of environmental, intrinsic and pre-and post-slaughter risk factors associated to dark-cutting beef in a Federal Inspected Type slaughter plant. Meat Sci. 2019:150:85-92. doi: 10.1016/j.meatsci.2018.12.007

[59] Garcia LG, Nicholson KL, Hoffman TW, Lawrence TE, Hale DS, Griffin DB, Savell JW, VanOverbeke DL, Morgan JB, Belk KE, Field TG, Scanga JA, Tatum JD, Smith GC. National Beef Quality Audit-2005: Survey of targeted cattle and carcass characteristics related to quality,

quantity, and value of fed steers and heifers. J. Anim. Sci. 2008:86:3533-3543. doi: 10.2527/jas.2007-0782

[60] FordyceG,WythesJR,ShorthoseWR, Underwood DW, Shepherd RK. Cattle temperaments in extensive beef herds in northern Queensland. 2. Effect of temperament on carcass and meat quality. Aust. J. Exp. Agric. 1988:28:689-693.

[61] Voisinet BD, Grandin T, Tatum JD, O'Connor SF, Struthers JJ. Feedlot cattle with calm temperaments have higher average daily gains than cattle with excitable temperaments. J. Anim. Sci. 1997:75:892-896. doi: doi: 10.2527/1997.754892x

[62] Francisco CL, Resende FD, Benatti JMB, Castilhos AM, Cooke RF, Jorge AM. Impacts of temperament on Nellore cattle: Physiological responses, feedlot performance, and carcass characteristics. J. Anim. Sci. 2015:93:5419-5429. doi: 10.2527/jas.2015-9411

[63] Cappellozza BI, Bastos JP, Cooke RF. Short communication: Administration of an appeasing substance to Bos indicus-influenced beef cattle improves performance after weaning and carcass pH. Livest Sci. 2020:238:104067. doi: 10.1016/j.livsci.2020.104067

[64] McGlone JJ, Anderson DL. Synthetic maternal pheromone stimulates feeding behavior and weight gain in weaned pigs. J. Anim. Sci. 2002:80:3179-3183. doi: 10.2527/2002.80123179x

[65] Archunan G, Rajanarayanan S, Karthikeyan K. Cattle pheromones. In: Mucignat-Caretta, C., editor. Neurobiology of chemical communication. Boca Raton (FL): CRC Press; 2014:461-488.

[66] Liberles SD. Mammalian pheromones. Annu. Rev. Physiol. 2014:76:151-175. doi: 10.1146/annurev-physiol-021113-170334

[67] Pageat P. 2001. Appeasing Pheromones to decrease stress, anxiety and aggressiveness. US Patent 6,054,481, 6,077,867, and 6,169,113 B1. January 2, 2001.

[68] Cooke RF, Millican A, Brandão AP, Schumaher TF, de Sousa OA, Castro T, Farias RS, Cappellozza BI. Short Communication: Administering an appeasing substance to Bos indicus-influenced beef cattle at weaning and feedlot entry. Animal 2020:14:566-569. doi: 10.1017/S1751731119002490

[69] Schubach KM, Cooke RF, Daigle CL, Brandão AP, Rett B, Ferreira VSM, Scatolin GN, Colombo EA, Pohler KG, Cappellozza BI. Administering an appeasing substance to beef calves at weaning to optimize productive and health responses during a 42-d preconditioning program. J. Anim. Sci. 2020:98:skaa269. doi: 10.1093/jas/skaa269

[70] Angeli B, Cappellozza BI, VasconcelosJLM,CookeRF.Administering an appeasing substance to Gir × Holstein female dairy calves on pre-weaning performance and disease incidence. Animals. 2020:10:1961. doi: 10.3390/ani10111961

[71] Colombo EA, Cooke RF, Brandão AP, Wiegand JB, Schubach KM, Duff GC, Gouvêa VN, Cappellozza BI. Administering an appeasing substance to optimize performance and health responses in feedlot receiving cattle. J. Anim. Sci. 2020:98:skaa399. doi: 10.1093/jas/skaa339

[72] Cappellozza BI, Cooke RF. Review: Effects of administration of an appeasing substance on performance, neuroendocrine stress response, and health of ruminants. Transl. Anim. Sci. (In review) TAS-2021-0864.

[73] Lawrie RA. Lawrie's Meat Science. 6th ed. Technomic Publishing Co., Lancaster, PA. 1988.

[74] USDA. United States Standards for Grades of Carcass Beef. United States Department of Agriculture, Agricultural Marketing Service, Livestock and Seed Division, Washington, DC. 1997.

[75] Hamilton DN, Ellis M, Miller KD, McKeith FK, Parrett DF. The effect of the Halothane and Rendement Napole genes on carcass and meat quality characteristics of pigs. J. Anim. Sci. 2000:78:2862-2867. doi: 10.2527/2000.78112862x

[76] van Laack RLJM, Stevens SG, Stalder KJ. The influence of ultimate pH and intramuscular fat content on pork tenderness and tenderization. J. Anim. Sci. 2001:79:392-397. doi: 10.2527/2001.792392x

[77] Wulf DM, Emnett RS, Leheska JM, Moeller SJ. Relationships among glycolytic potential, dark cutting (dark, firm, and dry) beef, and cooked beef palatability. J. Anim. Sci. 2002:80:1895-1903. doi: 10.2527/2002.8071895x

Chapter 12

Beef Consumption Pattern in Brazil

Eduardo Eugênio Spers, Pedro Carvalho Burnier and Thelma Lucchese-Cheung

Abstract

Brazil is one of the world's leading beef producers. The goal of this chapter is to give an overview of how Brazilian beef production is important to Brazil and worldwide. We also give an overview of some aspects of red meat consumption in this country and the main tendencies regarding sustainability production. The economic importance of beef production is Brazil is based on secondary data and the main content about beef consumption is based on researchs conducted by the authors that interviews Brazilian consumers. The chapter focus in some concepts, concerns and factors that affects consumption as symbolic aspects, ethical, health and environmental concerns, brand, herd tracking, guarantees of origin, legal employment, safety and hygiene, animal wellbeing, sustainability and the Carbon Neutral Beef initiative (CNB). Marketing, certification, treaceability and brand strategies conducted by some industries, the red meat premium boutiques in the retailing sector, and the role of brazilian government in meat safety, monitoring and regulation are also covered.

Keywords: Meat, Beef Production, Brazil, Consumer Behavior, Food Consumption

1. Introduction

Brazil is one of the world's leading beef producers not only as a consequence of the favorable environment to production and land available, but also as a result of decades of investment in technology that have increased not only productivity, but also the quality of the Brazilian product, making it competitive and allowing it to reach markets in more than 150 countries. Comparing with other meats, the volume of chicken meat exported in 2019 was 4.12 million tonnes, followed by 1.85 million tonnes of beef and 0.73 million tonnes of pork. However, sales with beef exports reached 7.57 billion dollars, followed by chicken with 6.90 billion and 1.58 billion in pork. The total for Brazilian meat exports accounted for 19.45% of the value of Brazilian agribusiness exports, with beef responsible 8.91% of that volume, which illustrates this sector's relevance to the Brazilian export agenda [1].

Forty years ago, the prevailing scenario of the Brazilian beef market was quite different. The herd was barely half the size of what it is now (213 million head), the focus was much more on supplying the domestic market, severe sanitation conditions blocked exports, degraded pastures predominated on the landscape and productivity was low. Over the last four decades cattle ranching has undergone a

IntechOpen

revolutionary modernisation process supported by technological advances in production systems and in organizing the supply chain that today are clearly reflected in the quality of Brazilian beef. The herd has more than doubled, while the pasture area has advanced little or has even diminished in some regions – a clear indication of increases in productivity. There has also been an increase in animal weight and reduction in mortality, increase in birth-rates and reduction of the time until slaughter. These gains have been possible thanks to the growing use of technologies by rural producers, especially along the lines of feeding, genetics and animal management and health.

In 2019 the Brazilian GDP was BRL 7.3 trillion, a nominal growth of 6.8% over the previous year. Part of that growth was due to the Ranching GDP, which registered a slight growth for the same period, illustrating the strength of the ranching sector in the Brazilian economy [1]. With a herd of approximately 215 million head, the Brazilian ranching sector in 2019 recorded 43.3 million head of cattle slaughtered. For that same period Brazil saw a 12.2% increase in beef exports that reached 2.49 million CWE. Of the total volume of meat produced, 76.3% or 8.01 million CWE went to the domestic market, while 23.6% were slated for export, the equivalent of 2.49 million CWE. Of the total exported, there was a 15.9% increase in the volume of fresh beef. That increase was due not only to the number of countries receiving exports, which went from 101 to 154, but also the increase in the volume of meat directed towards already consolidated markets such as China, where the volume of exports rose 54% from 2018 to 2019. During that same period the area of pastures utilized remained practically stable at 162.5 million hectares, with an average productivity that was also stable at 4.3 @/ha/year.

In 2019, Brazil ranks as the second largest beef producer worldwide, 10.49 million CWE, behind only the USA with 12.26, but ahead of Argentina 3.01. India with 2.91, and Australia 2.26. However, Brazil leads global exports with 2.49 million in CWE, followed by Australia with 1.56 and USA with 1,31 [1]. Among the importing countries, China stands out with a volume of 1.28; followed by the USA 1.30 and Hong Kong 0.388. It is worth pointing out that China received 50.4% of the volume of Brazilian exports in 2019. The European Union imported 3.05 million CWE, but only 180 thousand CWE from Brazil (5,9%). In terms of consumption, the USA stands out as the world's largest beef consumer 12.22 ml te c (37,1 kg/inhab. /year), followed by China 10.01 (7.18 kg/in hab./year); Brazil 8.06 (38.4 kg/in hab./year) and Argentina 2.29 (50.91 kg/inhab. year). One should note here the great potential for growth that China has, given that its consumption per inhabitant is only 18% of the Brazilian per capita consumption [1].

The goal of this chapter is to give an overview of how Brazilian beef production is important to Brazil and worldwide. We also give an overview of some aspects of beef consumption, marketing and retailing in this country. We finish presenting the main tendencies, most regarding sustainability production.

2. Consumption of red meat in Brazil

The symbolic aspect of food represents one of the factors that most influences the consumption of individuals. Proof of this symbolic weight can be observed in any empirical research, asking a consumer what food means to them. Da Matta [2] explained that the same food can have numerous meanings and such variation is dependent on the ways in which people think and judge such food. Through social interaction, the rules related to food and eating vary in time and space, being learned from an early age. Therefore, beliefs, values and emotions related to any food can say a lot about individuals' eating patterns and consumption behaviors.

The more a food, through its symbols, represents values considered important to consumers, the greater the chance of being chosen for consumption. Taking red meat as an example, in Western societies, its consumption is valued, for many consumer groups, for being associated with a status symbol, the strength for work and the guarantee of satiety [3–6].

On the other hand, in the context of contemporary food, there are also movements against the excessive intake of red meat consumption and, even, the justified suppression of animal protein, among others, for ethical, health and environmental reasons. Disregarding the proportionality between consumers and non-consumers of red meat but considering that the appreciation of the animal's intake, the reduction of consumption and the practice of abstaining from this act represent strategies of meat consumption. A better understanding of the determinants of such strategies can be clues to explain, for example, the transitions in current food models. For research on consumption behavior, it is important to access the ways in which consumers classify proteins (healthy x harmful, lean x fat, ordinary x festive, heavy x light) and decide to consume them (fresh x processed, chilled x frozen, meat in slab x cut, fresh x canned).

Red meat is a factor that simultaneously provides benefits and detrimental to health. Ranabhat Park and Kim [7] have conducted a study about the influence of alcohol and red meat Consumption on Life Expectancy in 164 Countries from 1992 to 2013 providing evidence on red meat consumption and life expectancy (LE) based on ecological analyses. Results suggest that high consumption of red meat has a negative impacto n LE in higher income countries (HIC and UMIC)[1]. Consumption of red meat over the acceptable level is positively associated with câncer. On the contrary, red meat consumption appears to have no influence on LE in lower income countries (LIC or LMIC). Animal source foods (ASF) such as red meat is still very important for developing countries from a nutritional point of view. Refraining from red meat consumption in these countries is not warranted. Under IGW categorization, Brazil fits into de UMIC (upper middle). In this case authors suggests that high- and middle- income countries are likely to revisit their policy regarding red meat in terms of production, sales, and consumption.

Other research on consumption intention also recognizes that individuals' ways of acting are determined by their value systems, their beliefs, their social rules and by taboos [8, 9]. Judgments guide the acceptance of a food, and its choice is conditioned by how it best represents a lifestyle, an activist stance, or conduct. In studies conducted in Belgium, Verbeke [8] proved that the greatest motivations for the purchase and consumption of food products were justified by their adaptations to an individuals' morals.

According to Vialles [10], the way individuals choose and prepare animals to be consumed is related to symbols, which determine a double strategy of meat preparation and consumption, zoophagy and sarcophagy. The first, without any change in shape, values the animal's presentation as it is, and can even be presented whole (for example, a fish or a piglet). The second proposes to mischaracterize the animal, valuing the consumption of its parts without being recognized (steaks, cuts of meat with culinary dishes such as stroganoff and sausages). Denying the death of the animal intended for human consumption and the difficulty in recognizing parts of an animal or the animal itself in one's food are sarcophagic characteristics.

With the purpose of measuring the sensitivity of consumers from different countries in Europe in relation to the consumption of red meat, Gautier [11] validated a scale of zoophagy (acceptance of the individual to recognize an animal as

[1] Using definitions from the World Bank International Gateway (IGW) as of January 2016, countries were categorized into the following 4 groups: Low-income countries (LIC), lower-middle-income countries (LMIC), upper-middle-income countries (UMIC), and high-income countries (HIC). Available online: http://data.worldbank.org/country

food) and sarcophagy (difficulty and malaise) the individual's ability to recognize an animal as food) by means of attitudinal and ideological determinants. On this scale, the individual expressed his value system, which was measured by personality indicators and other affective, symbolic and imaginary indicators of the attitude towards meat. Gautier's [11] zoophagy and sarcophagy scale was also tested by Cazes-Valette [3] in France. The results were interesting because consumption was not explained only by socioeconomic and demographic variables. The attitude and behavior towards meat explained, for example, the different points of view about the relationship between man and nature, about the ways of thinking and accepting the slaughter of animals for consumption, about the frequency of consumption, the ingested volume, and cutting and cooking preferences.

A study that examined animal wellbeing (AWB) was performed by Souza, Casotti and Lemme [12] seeking to understand reactions of consumers related to ill-treatment practiced against animals in industrial meat-producing processes which cause pain, suffering and stress. Their research shows that consumers generally are not aware of management standards in meat production and that 87% of respondents have difficulties connecting the food they consume with living animals. Even though meat is considered a commodity, some countries employ labelling schemes. The main criteria certified by those labels include herd tracking, guarantees of origin, management employed, safety and hygiene, animal wellbeing and so on.

The proposition of the beef acceptance index was useful to think about the attitudes of individuals classified in household surveys as strong consumers of animal protein in Brazil as it was shown on study carried by Lucchese-Cheung, T., Spers, E. E., Pereira, M. W. G. & Dias, P. C. S. P. [13]. An incredibly detailed analytical model was proposed, not only so that it could be replicated in other Brazilian states, but for determining other indexes that answer questions from applied research in other productive sectors. It was verified how the variables relate within the factors and among them, as well as to understand what they represent and, later, the determination of the independent variables on the attitudes of zoophagy and sarcophagy was measured.

The results are interesting when they point out statements of strong beef consumers that indicate a desire to reduce their consumption or replace it with white meat, or leaner cuts, for health reasons. In addition, food safety and certifications appear as signs of quality, which can be a tip for sector's agents to seek innovation in proposing quality seals and brands for the sector. On the other hand, more sarcophagic attitudes are described by feelings of pity, concern about the environmental problems that arise in the meat chain, as well as distrust in relation to the quality of the protein sold in retail.

Such attitudes were determined by feelings of emotion and the belief that man should protect nature. In general, ordinary beef consumers have many doubts about the production system of animals intended for human consumption, and it may be an opportunity for agents in the sector to invest in communication campaigns. The way food is perceived depends on the food culture of individuals that, in turn, guides their attitudes and behaviors. An index of zoophagy and sarcophagy was proposed in in this work, intending to measure the intensity of these attitudes among a group of consumers in the Brazilian region that declare itself as the largest consumer of fat meat in the country.

In light of theories that indicate that beliefs, social rules, emotions and personalities are as determinant of consumption as economic variables, this work found that the most zoophagic attitudes were determined by personality traits that reveal patterns of behavior that, in turn, indicate dissatisfaction and a need for always wanting more, in addition to extroversion and the desire to always be in a group and party.

The taste, the fat of the meat, the high frequency of consumption, the preference to see the meat hanging, and the fact that they think of animals as a food source for humans characterize zoophagic attitudes. The most emotional individuals, the smallest part of the interviewed group, sarcophagics, revealed negative emotions when talking about beef. They are consumers concerned with their choices because it has an impact for themselves and for others.

These reflections are original for marketing and consumer behavior studies, proving that the culture and symbolism of food play an important role in the way of acting and thinking of individuals. Thus, the agents of public and private power are interested in knowing that the taste for beef represents a strong cultural significance and that changes in consumption patterns represent fighting against feelings, emotions and their personal history.

Purchasing intentions motivated by ethical, health and environmental concerns have been identified as new drivers of consumption by a specific group of the population [14]. But, in general, what do Brazilian consumers think about the consumption of beef and the safety of the food they consume? What is the perception of these consumers in relation to the commercialized meat and the Brazilian production models? To what extent are your choices determined by human, animal and environmental health concerns?

The greater the perceived risk, the more consumers will seek alternatives to minimize it [15]. When it comes to the consumption of beef, the perception of risk can have an even greater impact on consumption decisions, since there is not always information that guarantees safety at the time of purchase and quality assessment is done during consumption [16, 17].

Investment in brand and certification attributes are examples of ways to guarantee access to relevant information in the attempt to provide greater security and generate trust with consumers [18]; Verbeke and Ward [19]. Thus, the certification of protein can be an attribute of choice of beef and a generating factor of greater confidence and perception of quality [20, 21]. Furthermore, on these attributes, Henchioni, McCarthy and Resconi [22] mention investments in guarantees of origin, in animal welfare, in pasture production systems with certificates of care with animal nutrition, guarantees of concerns with environmental issues, of traceability, genetic improvement and technologies that involve its processing.

Some research has been conducted on the consumption behavior and preferences of the beef consumer and the willingness to pay more for products with labels that attribute information on traceability and different types of meat quality guarantees [23–29]. In general, traceability can contribute to increasing consumer confidence in the entire food system as a means of attesting quality assurance [30]. Knowing about the animal's origin has been proven by studies as important information for European consumers [31, 32]. Certain countries are better regarded than others for their seriousness/credibility in food production. Origin is an important quality attribute that generates greater confidence. Consumer belief, on the other hand, is that quality/safety comes at a high price.

In a qualitative phase of the study conducted by Burnier [33] a clear distinction was identified in the involvement in a purchase depending on the type of event. According to one of the interviewed experts, there is a functional choice (day to day) and 'recreational' shopping for specific events (barbecue, dinner with friends). In the recreational choice there is a greater involvement with the product, a greater concern with the productive characteristics as well as quality of the meat. On the other hand, functional purchases are generally made by women and do not require specific information on the quality of the meat being purchased.

The study also indicates that involvement with the product appears as an important element in the discussion of sustainability in the meat sector. Another interviewed espert stated that the creation of the "meat academy" website exemplifies attention to meeting a real demand. Consumers have shown a greater interest in obtaining information on various attributes (besides softness) before or even after purchasing the product. The recent movement of 'gourmetization' in this market, evidenced in TV shows and movements by innovative chefs, increase consumer interest and consequently their greater involvement with the product.

The purchase occasion also influences the consumer in the definition of attributes that a product must have at the time of purchase and in the WTP for a higher quality and safer product. The day-to-day shopper has less concerns when compared to looking for a product for a special occasion (eg dinner with friends or a barbecue).

As a result of the focus groups, it was found that the group with a high level of involvement with the product showed concern with the meat production process, especially the AWE (animal wellfare). This interest was manifested lately in the participants since, at first, they did not consider aspects related to the animal, nor were they sensitive to socio-environmental attributes at the time of purchase. However, when presenting the different product choice options, AWE was a decisive factor for most of the group. This behavior was observed in both groups, being more evident in Group 1 (meat boutique), since they are more involved with the product.

3. Food safety and the red meat in Brazil

The guarantee of food quality is a growing focus of governments, companies and standardization agents. Efforts are addressed to control the attributes of a food product, with a peculiar care with the nutrition and safety characteristics. Guaranteeing quality is gaining notoriety as food consumption is being better appraised for the rulers, consumers and companies. This better evaluation suggests more voluntary quality assurances for companies and more regulation for government ([34], p. 409).

Food safety is consumer's warranty to acquire a food with quality and health attributes [35]. Institutions changes like new industrialization processes, new consumer demands, industrialization and urbanization, increase of competitiveness, development of scientific research, decrease of income expenses on food and globalization demands increasing the consumer, government and private organizations interest for safety and quality.

Meat is a source of protein and important vitamins and minerals. However, if produced and marketed in an incorrect way by private companies, can cause alimentary intoxication and, consequently, loss of value and reputation to final consumer. In the case of Brazilian beef the healthy profile are emphasized as a "green" cattle (that grows in the fields), different to those intensive processes that led to the mad cow disease that opens discussions about the State effectiveness in guaranteeing consumer's safety [36].

Understanding how consumer perception behaves over government enforcement policies and private branding strategies could help in an appropriate introduction of effectiveness communication over food safety aspects. Analyzing the process of meat purchase, Barcellos and Callegaro [37], interviewed 400 consumers in a Brazilian city and, through factor analysis, they reduced the variables on information quality indicators. Keeping constant the information about the animal, the product and choosing a specific type of meat, the consumer preference structure could be defined by the monitoring attributes (or public mechanism) and brand (private mechanism). We use these results as an assumption to the proposed model.

Scholars have discussed the level of knowledge that consumers have of the food production process within the boundaries of different constructs. Hanf and Kuhl [38] argued that quality, in consumer understanding, is a construct with multiple attributes, and they considered orientation throughout the process as one of the main dimensions of the quality control system. That is, the production system must be explicit: "from farm to fork". The authors noted that "providing traceability information and having a transparent production chain becomes a competitive necessity" ([38], p. 179).

The results found in a study conducted by Burnier et al. [39], involving 725 Brazilian consumers, suggest that concern with the production process is related to attitude and the intention to buy sustainable meat. The findings of this article confirm the importance of the "animal wellbeing" (AWB) and "traceability" attributes in the process of choosing meat. The results suggest that the meat industry and retail sector need to better explore the opportunity for differentiation at the moment of purchase. They can do this by increasing knowledge about carbon emissions and by disseminating the concept of animal wellbeing in formulating their communications strategies and in positioning products/brands that have socioenvironmental attributes offered to the final consumer. Traceability is another relevant attribute. The willingness to pay (WTP) for traceable meat is greater than that for paying for meat lacking traceability. The results of that study confirm the need for traceability in order to verify belief attributes. And food safety is an attribute very much in demand among consumers. The results make it clear that both types of traceability (back to the packing plant and back to the farm) are valued when compared to a non-traceable product. Participants in the study mentioned the presence of the Federal Inspection Service (SIF) seal as one of the items they observed at the moment of purchase. This confirms the importance of that seal because of its association with assurances of the meat's origin. In interviews with specialists, industry representatives indicated "farm to table" traceability as relevant to company operations as a means for validating food safety and fulfilling agreements reached with NGOs and the MPF.

4. The red meat industry in Brazil

The subject of branding has gained relevance in recent years in the field of agricultural commodities, particularly when the consumer is faced with making choices between similar products. By identifying reliable products, through known brands, with which they themselves identify, the consumer is able to make what they see as an advantageous purchase [38].

The joint actions of strong brands, at different levels of the production chain, can add value to the final product in terms of the consumer's perception of intangible attributes (such as food safety, traceability, and other attributes of trust) linked to the brand.

The production and industrialization sector of the food industry has gone through successive credibility crises due to product contamination, and so the notion of Food Safety has gained strength. Food safety has been the object of interest of several economic agents and some NGOs, who emerge as agents of pressure on the institutional environment, with the fear of a risk to their health down to the consumption of adulterated or contaminated foods.

Some surveys indicate that the food choices of consumers have been more influenced by concerns about the impact of food systems on human health - Food Safety. The perception of a food as safe appears to be a strong requirement in the choice of a product. Traceability during the different stages of the meat production chain is seen as a way of making the "quality" of the product more tangible [40].

As an example of response to this sort of pressure, in early 2013 a strong advertising campaign by the JBS company for its Friboi brand was launched as a means of making consumers aware of the importance of knowing the origin of the beef consumed in Brazil. This campaign was aired shortly after reports that highlighted the lack of controls at countless meatpacking plants in Brazil, casting doubt on guarantees of "quality"/soundness of the meat sold at retail points and butcher shops. Veja magazine published forceful articles on the issue, showing threats and diseases that Brazilians are subject to by consuming that type of meat coming from "clandestine" suppliers. These issues brought up an opportunity, nor only for sales, but of gaining a market share. "But for that to happen we had to deal with something very interesting and curious, which is people changing their habits. What we wanted, therefore, was for people to start asking by brand name for the meat they buy every day [...]", commented Marcio Oliveira, president of Lew' Lara\TBWA, the agency that came up with the Friboi campaign.

Grunert et al. [18] also emphasizes the importance of the brand as a way of minimizing consumer uncertainty at the time of purchase. The company can signal a product of superior quality, reduce the uncertainty of the consumer and encourage them to pay a premium price for superior quality [18].

The presence of the Federal Inspection Service (FIS) meat stamp was associated with the safety of the product by those interviewed in the Barcellos study (2007), while the certification stamps are associated with higher meat quality. The FIS stamp is usually present on the packaging and on the meat itself, meaning that it comes from animals that have been slaughtered in FIS-enabled slaughterhouses.

The brand appears as a relevant variable at the moment of buying meat. Respondents in the study by Burnier et al. [41] indicated a greater WTP for meat from a well-known and sustainable brand, which is in line with the findings of Tonsor and Shupp [42] and Grunert et al. [18]. However, unlike what was expected, the sustainable brand presented lower values than those of a well-known one. That indicates that consumers must have confidence in the brand they know because of intrinsic safety attributes instead of sustainability attributes. When it comes to extrinsic attributes, especially socioenvironmental characteristics, the brand will serve to identify the particularities of the productive process (breed, traceability, animal wellbeing, origin, etc.) that must be considered by more frequent consumers, especially those who make purchases day to day. In the qualitative study, interviewees indicated that, besides the tenderness attribute, they are concerned with knowing the animal's origin (traceability) and with the food's safety. The beef brand helps the consumer to have greater security in purchasing a product with the desired quality including a safer product obtained from controlled ranches.

The "safety" variable was included in the brand value construct and validated in this work that endorses that relation. The relevance of food security to the meat consumer appears in the work of Oliveira and Spers [40] and Hanf and Kuhl [38], where these researchers confirm the importance of production methods that have a format oriented towards guaranteeing traceability, so that they are capable of transmitting to consumers attributes of trust related to food security.

5. The red meat retailing sector in Brazil

The high-quality beef market in Brazil has changed greatly over the last 5 years and is becoming increasingly sophisticated. We can no longer call it a niche, since a short time ago meat packers were selling boxes to the major retail chains and now, they are selling truckloads. Premium and Gourmet beef have rigid production rules, including demands for animals with at least 50% European genetics. Meatpacking plants also prefer British breeds such as Aberdeen Angus and Hereford. Males must be young, less than 24 months old, castrated and with at most two permanent teeth, which characterizes a young animal. Females can have up to 4 teeth. The carcass must have a fat finish according to specifications from the meatpacking plants. The now, breeding cow will always be a Zebu, basically from the Nelore breed. The more it has been genetically improved, the better the cross will be, since the cow contributes 50% of the calf's quality. There are dozens of high-quality beef brands in Brazil. All are top of the line products from the major meatpackers and distributors. A large share is sold to steakhouses, restaurants and churrascarias, which look for grill cuts that meet high standards. Retail sales through supermarkets and butcher shops have been increasing constantly.

The Premium beef segment, unlike the one for wines and coffees, is not yet totally consolidated. Consumers do not yet have a clear perception of levels of quality such as differences in flavor, tenderness, texture and presentation of different and new cuts. A sizable number of consumers have reached a level where when they schedule a barbecue for friends, they no longer want to take risks in buying the meat. They prefer to pay a little more in order to be certain they are offering a quality barbecue [43].

Another tendency is the gourmet butcher shop. These establishments invest in different cuts for day-to-day recipes. The old-style butcher shops with tile flooring and hooks hanging from the ceilings have been a rarity in large urban centres for some time. A few survive, mainly in lower-income neighborhoods, but the place to buy beef now is officially at the supermarket. But supermarkets do not offer more sophisticated or newer cuts, or move away from the commonplace. And the growth in gastronomy means that consumer demands and the search for specialty meats have also increased. This is where the gourmet butcher shops come in, seeking to change the relationship between the client and the ingredient.

One of the pioneers in the business was Marcos Bassi; this chef was a butcher back in the 1960s. In his business he has always offered special cuts. His boutique store appeared in 1985 as an extension of a barbecue restaurant. It offers almost all of the cuts for a barbecue, as well as Argentine meats, game, craft beers and wines. This type of establishment is now in fashion, and the clientele continues to be faithful to the products, since it is possible to exchange a piece of meat that has not pleased them.

Another trend is for using the Angus line (a breed with superior quality meat) sold in packaged half-kilogram portions (always vacuum packed). That makes life easier for couples without children or even singles. For the visit to be complete, the store must also provide different types of salt, sauces and everything needed to complement the barbecue, in other words, an emporium completely focused on meat.

A common practice in this type of butcher's is for the owner to help the client to choose the cuts, provide feedback about appropriate quality and encourage the consumption of different products. Instead of the familiar picanha (rump steak) a cut traditionally favored by Brazilian barbecuers, some owners advise that it may be much more interesting to use other pieces such as shoulder (more marbled), bananinha (strips of meat from between the ribs, with a buttery taste) or, for those who prefer fillet mignon, the round cut (as tender as the fillet, but tastier). Besides these cuts being flavourful and appropriate for barbecue, they are also cheaper.

Additionally, it is important to discover the world that is behind the counter with its vacuum-packed red packages. The "world of beef" involves other factors, such as breeding the animals, the diet fed to the cattle, the time for slaughter... Each of these items will determine the type of product. Some of these entrepreneurs believe that the meat found in supermarkets is of doubtful quality. Consumers have less information about the procedence of what they are buying, when compared with the precedence good meat boutiques.

Another tendency is to promote an environment that people will enjoy visiting. Some entrepreneurs seek to closely control all stages of beef production and relate directly to the final consumer. Having classes and providing recipes at the sales point leads the customer to understand that "there is no second-rate meet, but instead poor preparation." It is thus possible to suggest less "noble" cuts (ground beef made from chuck, ragu made from knuckle or tender chuck) and stop the obssession Brazilians have of thinking that good meat is fillet mignon meat [44].

The butcher shop is being transformed in order to provide a more pleasant environment for the client, who is increasingly demanding, with a diversity of products and personalized service. That strategy seeks to strengthen smaller companies so that they can face competition from the supermarkets.

Butcher shops have seen that in order to survive they must offer differentiated services that add value to the product, says Manuel Henrique Faria Ramos, president of the Fresh Meat Retail Association in the State of São Paulo. That is no abstract tendency, but a reality, and those who do not follow it run the risk of having to close their doors, he states. According to Ramos, modernisation is necessary for butcher ships set up in more well-to-do neighborhoods, but is also important for establishments operating in less prosperous regions. According to the association, there are around 17 thousand butcher shops in the State of São Paulo. Of those, around 80% have incomes of up to BRL 240 thousand per annum, and only 2% go beyond BRL 2 million.

But to be a success, it is not enough to have a pretty shop; one must have the best products and the best services. Shop owners who do not modernize will become stuck in place. Consumers are more demanding. Besides direct sales to consumers, some stores organize courses that teach people how to barbecue. Specialists indicate that the movement towards modernizing butcher shop facilities is natural and similar to what happened earlier with bakeries: a movement to provide more comfort and wellbeing to the consumer. The butcher shop environment must be clean, but need not be uninviting, all in white [45].

6. Sustainable beef in Brazil

Sustainable consumption may be the result of a decision-making process that considers not only the individual needs of consumers (related to taste, price and convenience), but also perceptions and attitudes related to social responsibility (environment and fair trade), labelling (brand and seal) and sustainable food production (traceability, CO2 emissions and animal wellbeing). Sustainable products tend to be perceived by buyers and having better quality and greater social, environmental and economic value. The beef sector must recognize this consumer behavior in order to improve sustainability and to develop actions for meeting legislation related to that demand.

Food safety has become a point of interest to various economic agents, and the result is pressure coming from the institutional environment, because of the perception that there are probable risks to health due to consumption of adulterated or contaminated foods. The perception that a certain food is safe is now a strong requirement for differentiating its price. The origin of a product and the assurance of soundness during its production and sale processes have come to be valued by consumers. Traceability throughout the different stages of the chain can be a form of validating a products "quality".

A study by Imaflora "Comportamento e consumo verde dos brasileiros - o caminho para uma atuação social e ambientalmente responsável (Green behaviour and consumption among Brazilians – the path to socially and environmentally responsible action" [46] indicates that 51% of the interviewees are considered to have an "interested" profile. They are concerned with illegal deforestation, the destruction of native forests and the depletion of natural resources. They also have concerns that are reflected in their the day-to-day and consumption habits. Persons with this profile are beginning to inform themselves about sustainable practices, value local production and choose products that can be recycled. Water pollution, illegal deforestation and destruction of native forests are among the socioenvironmental issues that most concern Brazilian consumers. In the food sector the main factor in choosing foods is price, followed by Origin and Procedence (48%). The results of this study reinforce the importance of using seals and certificates to demonstrate socioenvironmental commitments, and help to direct communications activities that can promote and differentiate products and brands.

The beef sector is working to continuously improve its sustainability in order to achieve more desirable results in environmental, social and economic terms, which are increasingly more important for consumers. There are clearly observable efforts underway in the retail networks that encourage conscious consumption and that seek to help their clients think about how sustainability can become a part of the consumer's daily lives, by presenting them with initiatives they can participate in. In this case, with conscious consumption, consumers reflect on their purchasing habits and on the factors that determine their choices. They are provided conditions for analyzing the impact they can have on the ecosystem.

In this context, retail networks are demonstrating interest in selling products with socioenvironmental attributes; however, they emphasize that the lack of knowledge among final consumers makes it difficult to execute a policy of premium prices for "green" products. Since 2015, the three main retail networks in Brazil have been developing sustainable ranching platforms, demonstrating a strong commitment to monitoring the origin of the beef sold in their stores. Origin of the beef refers here to socioenvironmental practices by the ranches that produce the animals that will be slaughtered and offered for sale at the national retail level.

Sustainability in the beef sector is directly related to several socioenvironmental dimensions. Cattle ranching is considered to be one of the sectors that most contributes towards deforestation in Brazil. Conversion of forests into pasture is the best-known and documented impact coming from the beef production chain. Additionally, issues related to animal wellbeing, slave labour and emission of gases (CO2 and methane) appear as directly related to the expression "sustainable beef". In practice, the industry (meatpacking plants) and retailers also consider these topics as the basis for defining their sustainable ranching platforms. For these reasons, beef cattle ranching along with Brazilian soy are key commodities in international agreements in international agreements on climate changes [47].

In addition to the pressure exerted by demand, other actors have shown concern about the origin of the product and compliance with socioenvironmental attributes throughout the Beef supply chainbThe lack of an efficient traceability system in the Brazilian beef chain creates unnecessary risks that arise regarding the origin of cattle in the various links of the chain – e.g., meat processing industry, retailers, investors. Recently, international investors have questioned the major Brazilian meat producers, worldwide leaders in Beef production, about the origin of the animals that they slaughter in their plants in order to avoid running the risk of financing deforestation in Brazilian territory. Similar pressure is also being put on the beef production industry by retailers so as to avoid the risk of selling products obtained from areas with socioenvironmental problems (deforestation). Since 2009, the Federal Public Prosecution Service (MPF) has been promoting agreements with retailers and meatpacking companies covering all the links in the ranching chain that can guarantee that cattle are coming from areas without deforestation. Civil society has published studies that demonstrate the importance of monitoring and controlling cattle origin in order to assure that deforestation is suppressed throughout the beef production chain.

In response to these pressures the major Brazilian meat companies announced new platforms in late 2020 seeking to achieve efficient traceability and meet domestic and international demands so as to mitigate risks and promote transparency and environmental conservation. In August 2020, Marfrig, the second largest Brazilian meat exporter communicated its Marfrig Verde+ Plan that set out a programme for achieving a chain free of deforestation by 2030. In September it was the turn for another Brazilian producer, JBS, to communicate that it will monitor its entire supply chain to cut deforestation by 2025. That announcement was made weeks after the Norwegian asset management fund Nordea announced that it would withdraw its participation of some 40 million euros (around BRL 260 million) from JBS because of the company's lack of engagement in environmental issues. In 2019, after forest fires raged through the Amazon Rainforest, a group of 251 investors demanded a reduction in deforestation, identifying environmental impacts as "systemic risks" to their portfolios. Biodiversity and climate changes are important topics to agricultural markets that are exposed to extreme weather conditions, and to environmentally aware consumers, says Matt McLuckie, research director for Planet Tracker, a not-for-profit organization in the United Kingdom that seeks to redirect capital into sustainable development. "The trends have not been positive for agricultural producers, especially in the beef sector" [48].

7. Concluding remarks

The worldwide scenario for the beef chain favors an increase in Brazilian beef exports. Additionally, new markets are being opened, such as Russia, the Middle East and Asia. Therefore, the search for quality products and the satisfaction of needs and tastes among the different consumer markets is crucial for Brazil to remain an exporter of meats worldwide. Thus, Brazilian entrepreneurs from the ranching sector began to work with new concepts such as Traceability, Certification and Certifiers, beginning with institution of the Brazilian System for Registering Bovines and Bubalines (SISBOV), on 01/10/2002 [49].

The study of Burnier et al. [39] enables a better understanding of sustainable practices in the beef supply chain through identification and measurement of activities developed throughout the production process, which can help managers with formulating communication strategies and product/brand positioning in response to consumer concerns over the production process. These communication strategies should be a means of creating opportunities for more efficient modes of production, because they facilitate consumer understanding regarding responsible actions undertaken in the stages of the production process. Aware of consumer willingness to consume products with socio-environmental production attributes, the world's second-largest beef producing company launched the Carbon Neutral Beef initiative (CNB) in 2020, which responds to a call for productive efficiency, reduction of environmental impact and attention to animal welfare [50]. With this new CNB concept, several of the elements present in the scale proposed in this study are clearly communicated (animal welfare, traceability, environmental responsibility).

In Brazil, responding to the need to update traditional and inefficient production models, one can cite initiatives that advocate processes that are safer and valued for socio-environmental aspects such as the use of Integrated Crop-Livestock-Forest Systems (ILPF) for the production of meat, grains and wood. According to Alves, Almeida & Laura [51], among the advantages of these processes are sustainable intensification of land use, diversification of production, soil conservation, better use of natural resources and inputs, reduction of pressure by opening up new areas (earth-saving effect), animal welfare, carbon sequestration and the mitigation of gas emissions. Thus, the production of beef cattle, through systems in crop-livestock-forest integration (ILPF), makes Brazil an important player in the sector by offering sustainable meat produced in the tropics. On the adaptations that occurred in the production system, Embrapa Gado de Corte created a concept brand for beef, Carne Carbono Neutro (CCN). Alves, Almeida & Laura [51] explain that, among others, the objectives of the CCN concept brand are to offer quality guarantees for a certified product, derived from an innovative business model, concerned with carbon emissions, with the environment, with animal welfare, with the adoption of good agricultural practices and with the compliance with current Brazilian socio-environmental laws. A symbol of greater productive efficiency, reduction of environmental impact and concern for animal welfare, CCN represents a differentiated product that reflects adaptations in animal production systems to meet a new consumption trend.

A study by two Higher Education Institutions, UFMS and ESALQ/USP, in partnership with researchers from Embrapa Gado de Corte aimed to learn about the perception of Brazilians about the consumption of beef and about a Brazilian concept brand developed by Embrapa (Carne Carbono Neutro, or Neutral Carbon Meat). Attention should be paid to an important result obtained by the study. There is a representative consumption behavior in Brazil, attentive to good production practices, in addition to positive attitudes related to the initiatives for presenting innovative products to the market that respond to consumer concerns related to animal welfare and the association of attentive food production. Legal bases and sustainability issues. On the other hand, at least a quarter of the interviewed population is represented by attitudes of indifference in relation to production issues and, also, of denial and disbelief in relation to the CCN concept brand. The variables that best contributed to explain this type of behavior are related to the lack of information or knowledge and the notion of risk. The lack of information or the erroneous information that reaches the final consumer about the problems related to food safety and the risk of consumption represent barriers to the commercialization of conventional protein and, mainly, to the CCN concept brand since it is an innovative product in the market. This study found that innovation was a generating factor of distrust on the part of final consumers. Thus, considering the future challenges for food supply, the research carried out in the national territory and analysis of the proposed typology allows us to state that there is an agenda of opportunities for productive systems that value socio-environmental aspects and responsible production. However, these initiatives will only make sense to the consumer population with a lot of investment in communication campaigns. Being assertive, the availability of most consumers to consume and pay for the differentiated product will be verified.

Another tendency now appearing in Brazil are meat boutiques with premium, gourmet and in some cases sustainable products. Beef Passion is a brand of beef produced in house, that is certified and sustainable and offers 72 exclusive cuts. The company prioritizes animal wellbeing through management that conditions the animals to being docile based on trust between different species, including humans, until the mature age for slaughter. On a daily basis, they use methods for conditioning the animals, such as ambient sound in the SPA and the sound of a cow horn during handling, as well as interaction with balls and people so that they can adapt to different situations in which they will pass through the next stages of their lives. Along with that the team is trained in methods for leading the animals, always with great calm and respect. The Beef Passion team believes that cattle are sacred animals that must be treated throughout their entire life cycles. All the production is tracked, raising is done in under an extensive system in central western Brazil, fattening occurs in a system with intensive supplments and pastures and finishing is done at the bovine SPA in the state of São Paulo. Genetic selection seeks to optimize the fat and fibers in the carcass. The company crosses the Angus and Wagyu breeds originally fron Australia and Japan, producing cuts under the "Australian Passion" and "Grand Passion" seals. After stunning and slaughter, the carcasses rest for 48 hours before being portioned and vacuum-packed. Boneless cuts are chilled, while those with bones are frozen according to the federal inspection norms (SIF) for Brazil.

As a result, the company has a standardized product of high quality and low levels of saturated fat, with 70% of the fat in the meat being unsaturated. Futhermore, the meat is certified as 100% sustainable by Rainforest Alliance, which attests to socioenvironmental excellence throughtout the production system [52].

Author details

Eduardo Eugênio Spers[1*], Pedro Carvalho Burnier[2] and Thelma Lucchese-Cheung[3]

1 Departament of Economics, Administration and Sociology (LES), Luiz de Queiroz College of Agriculture (ESALQ), University of São Paulo (USP), Brazil

2 ESPM, PPGA, São Paulo, Brazil

3 School of Business and Administration, Federal University of Mato Grosso do Sul (UFMS), Brazil

*Address all correspondence to: edespers@usp.br

References

[1] ABIEC. (2020). *Relatório Anual 2020. Perfil da Pecuária no Brasil. Disponível em:* <http://abiec.com.br/publicacoes/beef-report-2020/>.Accessed on 12 January 2021.

[2] DA MATTA, R. The que faz o Brazil, Brazil? Rio de Janeiro: Rocco, 1984.

[3] Cazes-Valette G. Genres et viandes: vers un troisième sexe. Sociologie et anthropologie de l'alimentation, 2005.

[4] BARROS, G. S. D., MENESES, J. N. C., & SILVA, J. A. D. Representações sociais do consumo de carne in Belo Horizonte. Physis: Revista de Saúde Coletiva, 2012, 22, 365-383.

[5] LUCCHESE-CHEUNG, T.; BATALHA, M. The.; LAMBERT JL. Comportamentos do consumidor de alimentos: tipologia e representação da comida. Agroalimentaria, v. 18, n. 35, 2012.

[6] FARGE, S.; MORETTI, S. L'Imaginaire Culinaire En Allemand, Espagnol Et Français: Le Rapport À La Viande (The Alimentary Imaginary in German, Spanish and French: How Meat is Experience). ESSACHESS-Journal for Communication studies, v. 8, n. 2, p. 13-25, 2015.

[7] Ranabhat Cl, Park MB, Kim CB. Influence of Alcohol and Red Meat Consumption on Life Expectancy: Results of 164 Countries from 1992 to 2013. Nutrients 2020, 12, 459; doi: 10.3390/nu12020459

[8] VERBEKE, W.; Marcu, A.; Rutsaert, P.; Gaspar, R.; Seibt, B.; Fletcher, D.; Barnett, J. 'Would you eat cultured meat?': Consumers' reactions and attitude formation in Belgium, Portugal and the United Kingdom. Meat science, v. 102, p. 49-58, 2015.

[9] HUNG, Y.; DE KOK, T. M.; VERBEKE, W. Consumer attitude and purchase intention towards processed meat products with natural compounds and a reduced level of nitrite. Meat science, v. 121, p. 119-126, 2016.

[10] Vialles N. (1987). Le sang et la chair. Paris: Ed. de la maison des sciences de l'homme (coll. Ethnologie de France).

[11] GAUTIER, J.-M.; D'AIDE À LA DÉCISION, S. I. A. D. Psychologie du consommateur et comportement d'achat: mise en place et validation d'une échelle de personnalité. Groupe HEC, 2001.

[12] SOUZA, M. C., CASOTTI, L., & LEMME, C. (2013). Consumo consciente como determinante da sustentabilidade empresarial: respeitar os animais pode ser um bom negócio? Rev. Adm. UFSM, Santa Maria, v. 6, Edição Especial, p. 229-246, maio.

[13] LUCCHESE-CHEUNG, T., SPERS, E. E., PEREIRA, M. W. G. & DIAS, P. C. S. P. (2021). Beef acceptance index proposition. Revista de Economia e Sociologia Rural, 59(2), e223297. https://doi.org/10.1590/1806-9479.2021.223297

[14] LIU, R., HOEFKENS, C., & VERBEKE, W. (2015). Chinese consumers' understanding and use of a food nutrition label and their determinants. Food Quality and Preference, 41, 103-111.

[15] PENNINGS, J. M., WANSINK, B., & MEULENBERG, M. T. (2002). A note on modeling consumer reactions to a crisis: The case of the mad cow disease. International Journal of research in marketing, 19(1), 91-100.

[16] ANGULO, A. M., & GIL, J. M. (2007). Risk perception and consumer willingness to pay for certified beef in Spain. Food Quality and Preference, 18(8), 1106-1117.

[17] Spiterni Cornish L and Moraes C. (2015). The impact of consumer confusion on nutrition literacy and subsequent dietary behaviour. Psychology & Marketing, 32(5), 558-574.

[18] GRUNERT, K. G., BREDAHL, L., & BRUNSO, K. (2004). Consumer perception of meat quality and implications for product development in the meat sector—a review. Meat science, 66(2), 259-272.

[19] VERBEKE, W., & WARD, R. W. (2006). Consumer interest in information cues denoting quality, traceability and origin: An application of ordered probit models to beef labels. Food quality and preference, 17(6), 453-467.

[20] APRILE, M. C., CAPUTO, V., & NAYGA JR, R. M. (2012). Consumers' valuation of food quality labels: the case of the European geographic indication and organic farming labels. International Journal of Consumer Studies, 36(2), 158-165.

[21] PARDO, M.A., JIMÉNEZ, E. and PÉREZ-VILLAREAL, B. (2015), "Misdescription incidents in seafood sector", Food Control, Vol. 62 No. 11, pp. 277-283.

[22] HENCHIONNI, M.M., McCARTHY, M. and RESCONI, V.C. (2017), "Beef quality attributes: a systematic review of consumer perspectives", Meat Science, Vol. 128 No. 1, pp. 1-7.

[23] CICIA, G., & COLANTUONI, F. (2010). Willingness to pay for traceable meat attributes: a meta-analysis. International Journal on Food System Dynamics, 1(3), 252-263.

[24] DICKINSON, D. L., & VON BAILEY, D. (2005). Experimental evidence on willingness to pay for red meat traceability in the United States, Canada, the United Kingdom, and Japan. Journal of Agricultural and Applied Economics, 37(3), 537-548.

[25] HOBBS, J. E., BAILEY, D., DICKINSON, D. L., & HAGHIRI, M. (2005). Traceability in the Canadian red meat sector: do consumers care?. Canadian Journal of Agricultural Economics/Revue canadienne d'agroeconomie, 53(1), 47-65.

[26] LIM, K. H., Hu W, Maynard LJ & Goddard E. (2014). A taste for safer beef? How much does consumers' perceived risk influence willingness to pay for country-of-origin labeled beef. Agribusiness, 30(1), 17-30.

[27] VAN RIJSWIJK, W., & FREWER, L. J. (2012). Consumer needs and requirements for food and ingredient traceability information. International Journal of Consumer Studies, 36(3), 282-290.

[28] VAN RIJSWIJK, W., FREWER, L. J., MENOZZI, D., & FAIOLI, G. (2008). Consumer perceptions of traceability: A cross-national comparison of the associated benefits. Food Quality and Preference, 19(5), 452-464.

[29] MENOZZI, D., HALAWANY-DARSON, R., MORA, C., & GIRAUD, G. (2015). Motives towards traceable food choice: A comparison between French and Italian consumers. Food Control, 49, 40-48.

[30] Hobbs JE. (2016). Effective Use of Food Traceability in Meat Supply Chains. In Advances in Food Traceability Techniques and Technologies (pp. 321-335). Woodhead Publishing.

[31] TELLIGMAN, A. L., WOROSZ, M. R., & BRATCHER, C. L. (2017). A qualitative study of Southern US consumers' top of the mind beliefs about the safety of local beef. Appetite, 109, 1-10.

[32] SPENCE, M., STANCU, V., ELLIOTT, C. T., & DEAN, M. (2018). Exploring consumer purchase intentions towards traceable minced beef and beef steak using the theory of planned behaviour. Food Control, 91, 138-147.

[33] Burnier PC . A influência da dimensão ambiental na atitude, na intenção de compra e no desejo de pagar pela carne bovina. 177 p. Tese (Doutorado em Administração) – Escola Superior de Propaganda e Marketing, São Paulo, 2018.

[34] CASWELL JA. Valuing the benefits and costs of improved food safety and nutrition. The Australian Journal of Agricultural and Resource Economics, vol 42, no. 4, p. 409-424, 1998.

[35] SPERS, Eduardo Eugênio. A Segurança alimentar ao longo da cadeia. Conjuntura Alimentos, v. 5, n. 1, p. 18-26, fev. 1993.

[36] Enriquez-Cabot; Goldberg RA. Technology Crises and the Future of Agribusiness: BSE in Europe. Case Study: Harvard: Harvard Business School, 1996, 41p.

[37] de Barcellos MD; CALLEGARO, C. A. M. A Importância da informação como indicador de qualidade: o caso da compra de carne bovina em Porto Alegre, Anais: Encontro Nacional de Pós-Graduação em Administração (ENANPAD), Salvador, 2002, 15p.

[38] HANF, J.H., & KÜHL, R., (2005). Branding and its consequences for German agribusiness. Agribusiness 21(2), 177-189. https://doi.org/10.1002/agr.20042.

[39] BURNIER PC; GUERRA, D.; SPERS, E. E. "Measuring consumer perceptions over beef good practices and sustainable production process", British Food Journal, Vol. ahead-of-print No. ahead-of-print, 2020. https://doi.org/10.1108/BFJ-12-2019-0904

[40] OLIVEIRA, R., & SPERS, E. (2018). Brand equity in agribusiness: Brazilian consumer perceptions of pork Products. *Revista de Administração de Empresas*, 58, (Jul.-Aug.).

[41] Burnier, P. C., Spers, E. E.; de Barcellos, M. D. (2021). Role of sustainability attributes and occasion matters in determining consumers' beef choice. Food Quality and Preference, vol 88, 104075. https://doi.org/10.1016/j.foodqual.2020.104075

[42] TONSOR, G, & SHUPP, R. (2009). Valuations of 'Sustainably Produced' Labels on Beef, Tomato, and Apple Products. Agricultural and Resource Economics Review, 38/3, 371-383.

[43] GAZETA DIGITAL, 2018 (https://www.gazetadigital.com.br/colunas-e-opiniao/colunas-e-artigos/a-evolucao-da-carne/544039) accessed on 6 February 2021

[44] NOSSA COZINHA, 2014 (https://www.uol.com.br/nossa/cozinha/noticias/redacao/2014/08/01/acougues-gourmet-apostam-in-cortes-diferentes-para-receitas-do-dia-a-dia.htm) accessed 6 february 2021

[45] BEEF POINT, 2013 (https://www.beefpoint.com.br/acougues-brasileiros-acompanham-as-novas-tendencias-e-se-transformam-in-boutiques-de-carne/) accessed 01 february 2021

[46] IMAFLORA, Comportamento e consumo verde dos brasileiros - o caminho para uma atuação social e ambientalmente responsável, 2019 https://www.imaflora.org/public/media/biblioteca/5dad9aaf1b5a1_pesquisa_final.pdf accssed 10 january 2021

[47] TROPICAL FOREST ALLIANCE 2020. Impacts of supply chain commitments on the forest frontier. 6 Jun. 2018. Disponível em: <https://www.tfa2020.org/wp-content/uploads/

2018/06/Impacts-of-Supply-Chain-Commitments-on-the-Forest-Frontier.pdf> . Acesso em: 6 jun. 2018.

[48] CARTA CAPITAL 2020 https://www.cartacapital.com.br/sustentabilidade/investidores-procuram-alternativas-a-industria-da-carne/ accessed on 10 January 2021

[49] BEEF POINT, 2004 https://www.beefpoint.com.br/certificacao-da-carne-bovina-20293/) accessed on 15 January 2021

[50] Villas Alves , F. et al . (2015) Carne Carbono Neutro: um novo conceito para carne sustentável produzida nos trópicos [recurso eletrônico] - Campo Grande, MS : Embrapa Gado de Corte. 29 p. ; 21cm. - (Documentos / Embrapa Gado de Corte, ISSN 1983-974X ; 210)

[51] ALVES, F. V., Porfírio-da-Silva V, & Karvatte Junior N. (2019). Bem-estar animal e ambiência na ILPF. Embrapa Gado de Corte-Capítulo in livro científico (ALICE).

[52] BEEF PASSION https://www.beefpassion.com.br/ accessed on 10 February 2021